Bird Feather Science

Bird Feather Science

Authors

Dr. (Mrs.) Rucha S. Karanjekar
M.Sc., Ph.D.
Department of Zoology,
Sevadal Mahila Mahavidyalaya,
Nagpur – 440 009 (M.S.), India

&

Prof. Pravin N. Charde
M.Sc. (Zoo.), Ph.D., FICER
Principal,
Sevadal Mahila Mahavidyalaya,
Nagpur – 440 009 (M.S.), India

2013

BIOTECH BOOKS®

ISBN 978-81-7622-275-4

Published by: **BIOTECH BOOKS®**
4762-63/23, Ansari Road, Darya Ganj,
New Delhi - 110 002
Phone: +91-011-23262132
E-mail: biotechbooks@yahoo.co.in

Printed at: **Chawla Offset Printers**
Delhi - 110 052

PRINTED IN INDIA

Preface

The basic structure of feather is described in this book. And a new logic of feather science is applied to study the feathers. This book reveals the application of a new and innovative concept called ptilochronology to study the sexual behavior of birds specially Japanese Quail. Ptilochronology is an interesting tool by which feather growth can be measured. The idea of ptilochronology was first introduced by Grubb in 1989. It is a growing branch of ornithology which opens ways to a number of areas such as avian behavior, ecology, conservation biology and remains a lot to do. This is the first study on this subject in the country.

Our aim is to share this innovative technique with the developing researchers in science and veterinary science, a new concept with the students, an interesting idea with the bird lovers, bird watchers and knowledge with the other readers. As you will read this book, you will learn how to use the technique.

Acknowledgements are made to the publishers for publishing this book. We feel that this book will definitely be an addition to your knowledge. Thank you for reading the book.

Dr. (Mrs.) Rucha S. Karanjekar

Prof. Pravin N. Charde

Preface

Contents

Chapter 1
Feather:
Structure and Arrangement

Feather Topography

A typical wing feather consists of a central, stiff shaft with the softer vanes on each side. The leading edge of the feather during flight is called the **outer vane**. The opposite vane is wider than the outer vane and is referred to as the **inner vane** (Figure 1.1).

Feathers consist of the following structural elements.

Central Shaft

The central shaft of a feather is divided into two regions.

The **calamus** is the part of the shaft closest to the bird's body. It is hollow and does not contain any vanes.

The distal end of the central shaft is referred to as the **rachis**. The rachis is solid and is defined as the area to which vanes are attached.

Vanes

The vanes extend from each side of the feather. A series of parallel branches called **barbs** make up the vane.

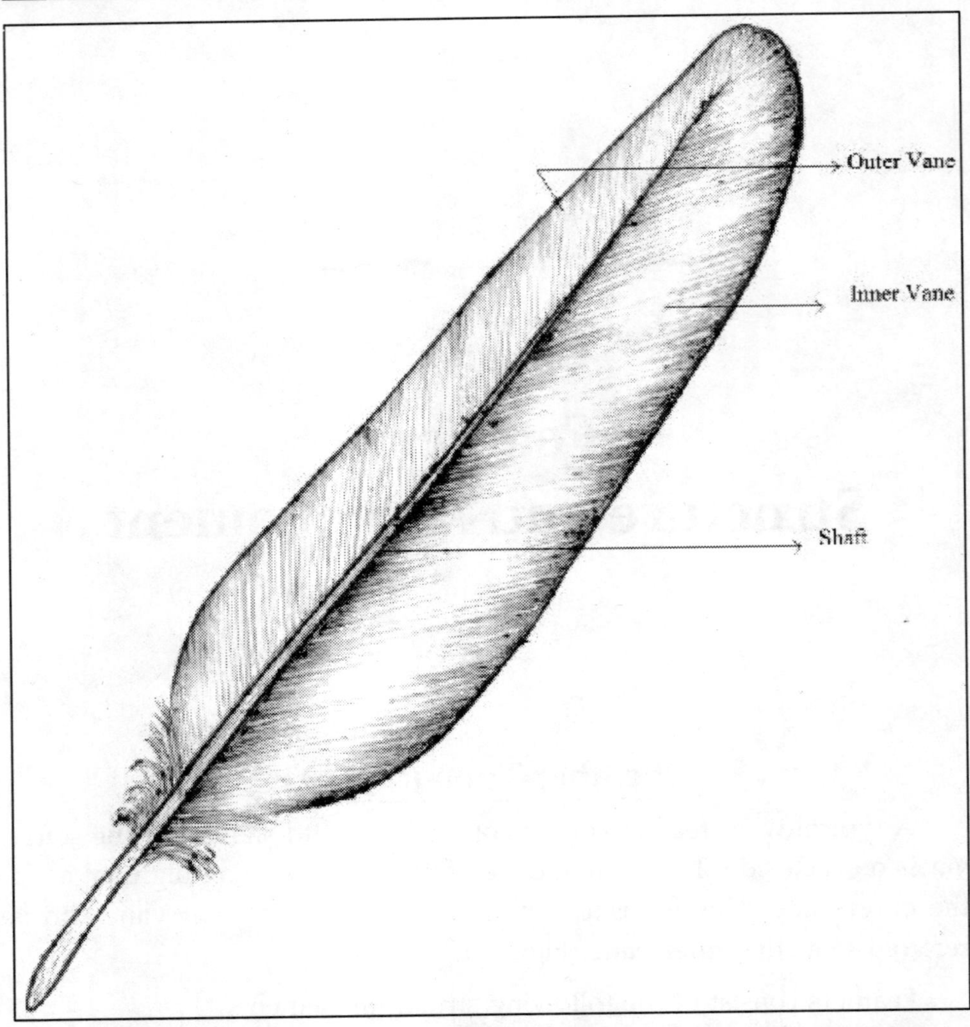

Figure 1.1: Basic Structure of Typical Wing Feather: Stiff Central Shaft Runs Entire Length of Feather, with **Vanes** Extending to Either Side. Note asymmetry in vane width, with narrower vane on edge of wing that leads in flight – termed the **outer vane**. The wider vane is called the **inner vane**.

Extending from the barbs are a series of short branchlets called **barbules**. Tiny hooklets tie the barbules, and ultimately the barbs, together. This somewhat complex arrangement creates the strong but light structure of the feather.

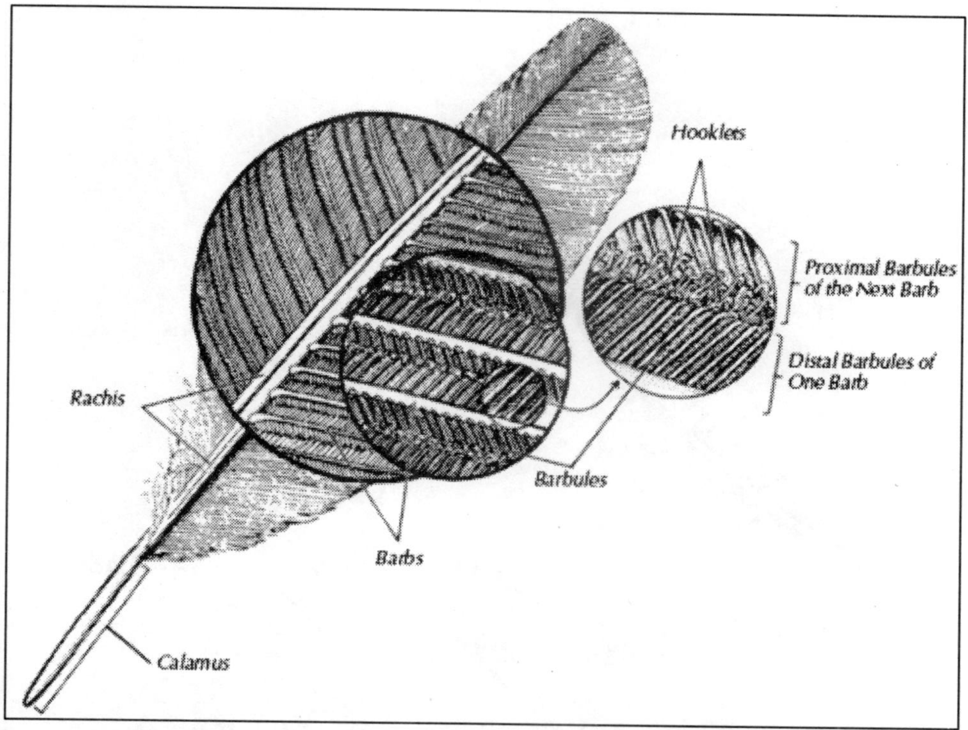

Figure 1.2: Basic Structure of Typical Wing Feather Showing Arrangement of Barbs and Barbules.

Feather Tracts

Feathers are not attached to birds in a random manner over the entire body of the bird. Instead they are usually found in often linear tracts celled **pterylae**. The spaces on the bird's body without feather tracts are referred to as **apteria**. The densest area for feathers is often on the bird's head and neck.

Types of Feathers

Contour Feathers

When you look at a bird the contour feathers are the outermost feathers, or the ones you see. They provide the color and the shape of the bird. The wing feathers are strong and stiff, supporting the bird during flight. The contour feathers tend to lie on top of each other, much like shingles on a roof. The feathers therefore tend to shed rain, keeping the body dry and well insulated.

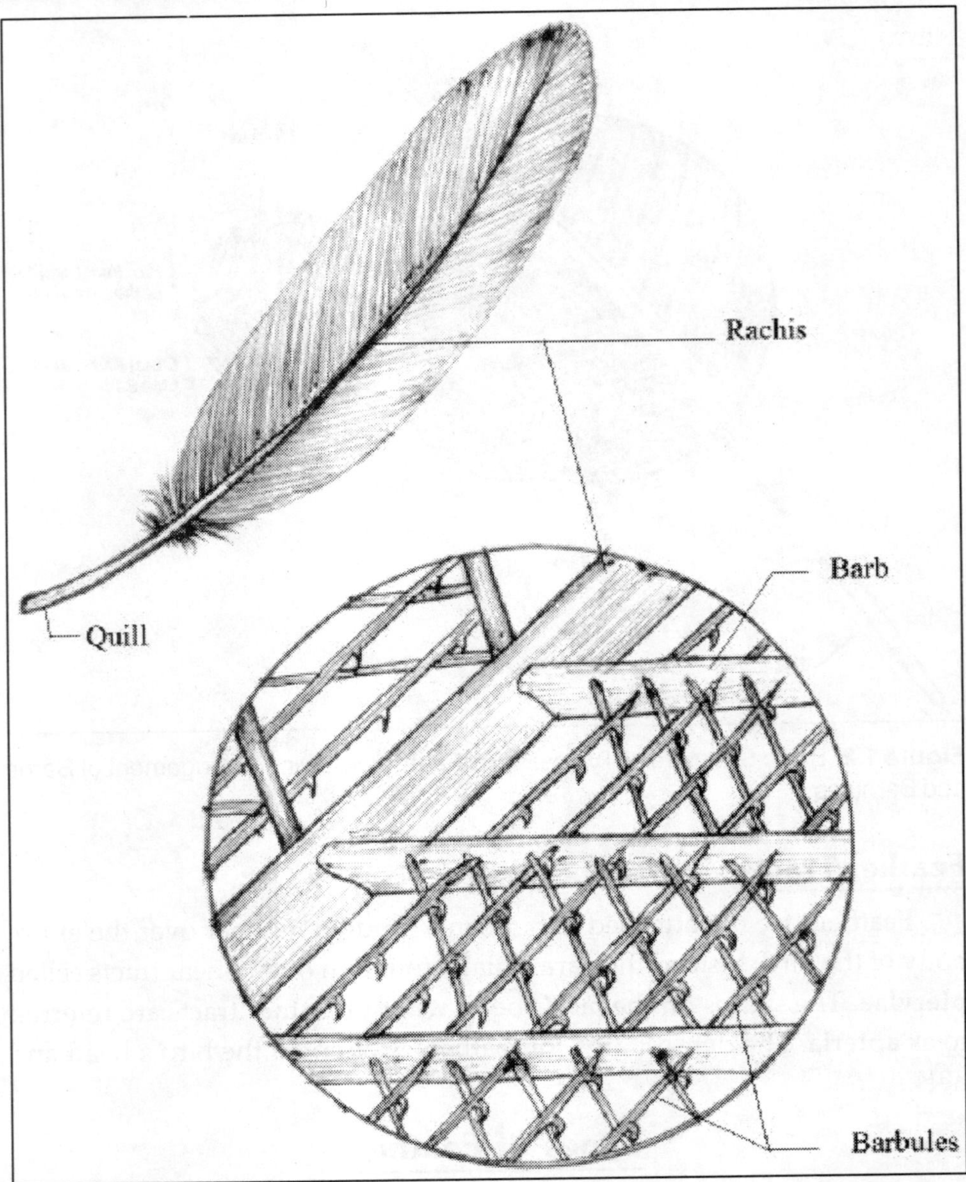

Figure 1.3: Basic Structure of Wing Feather Showing Arrangement of Barbs and Barbules.

Each contour feather can be controlled by a set of specialized muscles which control the position of the feathers, allowing the bird to keep the feathers in a smooth and neat condition.

Remiges

The largest contour feathers are often the large flight feathers, which are collectively called the **remiges**. Since they are responsible for supporting the bird during flight, remiges are attached by ligaments or directly to the bone. The outer remiges are referred to as the **primaries** and are the largest and strongest of the flight feathers. They are attached to the skeletal equivalent of the "hand" of the bird.

The inner remiges are called the **secondaries** and are attached to the "forearm" of the bird. They are located between the body of the bird and the primaries. The secondaries provide lift in both soaring and flapping flight.

Rectrices

The tail feathers are used to provide stability and control. They are referred to collectively as **rectrices**. The rectrices are connected to each other by ligaments, with only the innermost feathers attaching directly to the tailbone.

Coverts

Bordering and overlaying the edges of the remiges and the rectrices on both the lower side and upper side of the body are rows of feathers called **coverts**. The coverts help streamline the shape of the wings and tail while providing the bird with insulation.

Afterfeather

Attached to the lower shaft of some contour feathers are the typically much smaller **afterfeathers**. The afterfeathers resemble the main feather and provide an extra layer of warmth.

In North American birds, the afterfeathers of grouse are especially well developed for their life in seasonally cold and arctic regions.

The flightless Emu of Australia has specially adapted after feathers that are as large as the main feather and provide protection as the bird moves through the thick brush of its natural habitat. Relatively recent additions to the Emu's normal range are four-foot high sheep fences topped with barbed wire. Seemingly unperturbed, the flightless Emu crosses these fences by running straight into them, resulting in a high-speed somersault over the fence. The tuft of feathers left behind is a sure indication of an Emu crossing and a testament to the amount of protection the feathers offer.

Bristles

Highly specialized feathers, **bristles** are small contour feathers which lack barbs on the outermost part and have an especially stiff rachis.

Rictal bristles project from the beak of many insect-eating birds, including flycatchers, nightjars and even the American Robin. They are believed to provide protection for the bird's eyes as it consumes its wriggly prey. The bristles may also provide tactile feedback, like the whiskers on a dog or cat.

Feather Arrangement

Feathers are more or less arranged in a typical pattern on a bird's wing. The Figure 1.4 shows the diagrammatic representation of arrangement of feathers on a wing. The feathers that make up the typical bird's wing are:

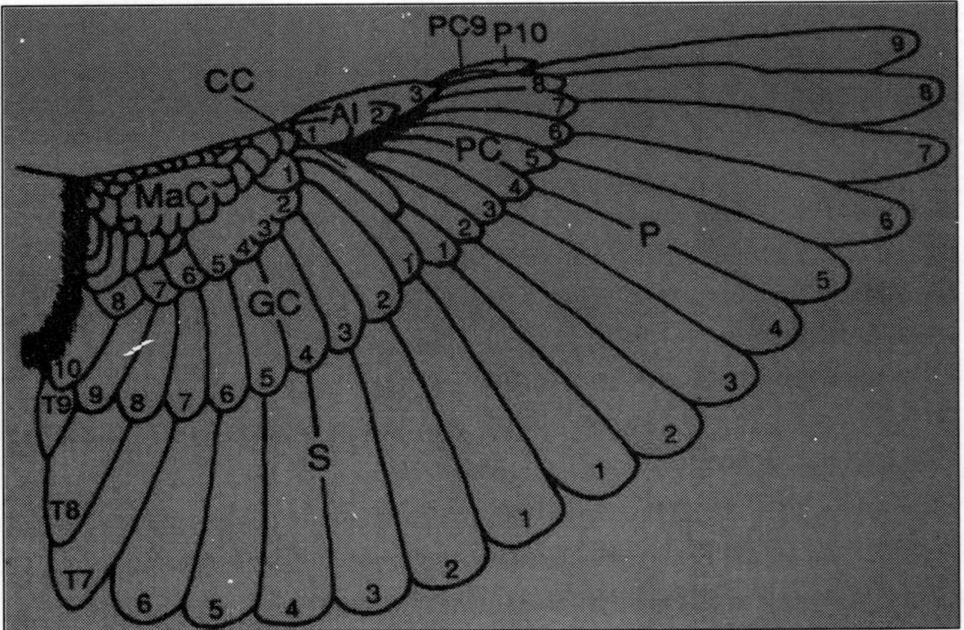

Figure 1.4: Diagrammatic Representation of Feather Arrangement on the Wing Showing MaC-marginal coverts, MeC-median coverts (1-8), GC-greater coverts (1-10), CC-carpal covert, Al-alula feathers (1-3), PC-primary coverts (1-9), Tertials-(T7,T8,T9), S-Secondaries (1-6), P-Primaries (1-10).

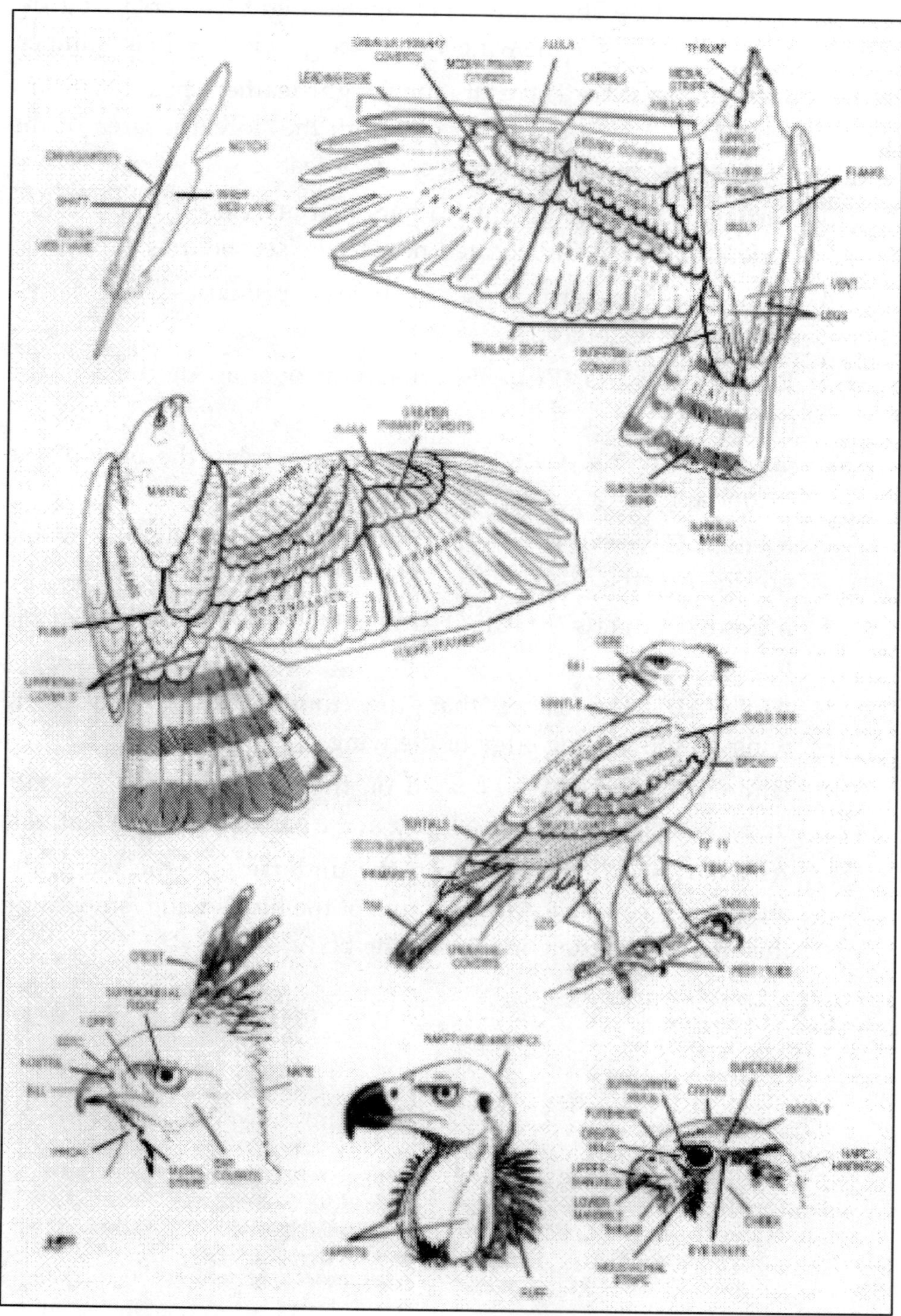

Figure 1.5: Figure Showing Topography of a Raptor.

1. **Primaries (P)** – These are the elongated flight feathers that grow out from the end of the wings. They are generally 9-10 in number.
2. **Secondaries (S)** – These are long flight feathers positioned just behind primaries and grow out from the 'forearm' area of the wing. Many birds have six secondary feathers.
3. **Tertials (T)** – Three flight feathers that are closest to the bird's body along the wing, located next to the secondaries.
4. **Remiges** – This term is used to refer to primaries, secondaries, and tertials together.
5. **Primary Coverts (PC)** – Feathers that overlap the base of the primaries.
6. **Greater Coverts (GC)** – Feathers that overlap the base of the secondaries.
7. **Median Coverts (MeC)** – Feathers that overlap the base of the greater coverts.
8. **Marginal Coverts (MaC)** – Feathers that overlap the base of the median coverts.
9. **Alula (Al)** – The feathers that grow from the 'thumb' area of the wing on the leading edge of the wing.
10. **Underwing coverts** – Located on the underside of the wing, underwing coverts create a lining at the base of the flight feathers.
11. **Auxillaries** – Also located on the underside of the wing, the auxilliaries cover the 'armpit' area of the bird's wing, smoothing the area where the wing meets the body

Chapter 2
Concept of Ptilochronology

Ptilochronology (Grubb 1987) has been proved to be an effective tool to know the overall nutritional condition of the birds. It is found to be a useful measure in characterizing the responses of birds to various environmental factors, both harmful and beneficial (Grubb 2006). A fundamental assumption behind the logic of ptilochronology is that there is a positive relation between the nutritional condition of a bird, as indexed by the growth rate of its feathers, and its biological fitness (Grubb 1989). The method is a non-invasive technique wherein by plucking the central rectrix the birds are allowed to grow a replacement feather, also known as the 'induced feather'. The induced feather is also plucked for evaluating data (*i.e.* average daily growth bar width, number of fault bars, total length). The feather comprises alternate light and dark bands. A pair of consecutive dark and light bands represents 24 hours of growth, termed as a 'growth bar' (Michener and Michener 1938); which represents the nutritional condition of the bird during those 24hrs. of feather growth (Wood 1950) (Plate 2). Sometimes there appear translucent bands produced by stressful events during feather formation called 'fault bars' (Plate 3). They weaken feathers and increase their probability of breakage, and thus could compromise bird fitness by lowering flight performance (Serrano *et al.,* 2005).

Plate 1: A Pair of Adult Male and Female Japanese Quail (*Coturnix coturnix japonica*).

Male characteristics are considered to be an honest signal of male quality, that (signals) can be used by the females for mate selection. Ptilochronology (growth bar width) has proved to be a potentially honest signal of bird quality (here 'quality' refers to ability to confer genes to the next generation by attracting one or more mates per breeding season, by surviving for many breeding seasons, or both; Grubb 2006). Growth bar width in males is positively correlated with individual quality and hence total reproductive success in a season (Takaki *et al.*, 2001).

In this book, we are implementing ptilochronology (Grubb 1987) to study sexual selection in Japanese Quail. Females use multiple signals or cues for mate choice (Candolin 2003). Males display individual quality through their ornamentation depending on their body condition, which facilitates the females to make mate choice (Zahavi 1975). Bird plumage is useful to study the correlation between the body condition and behavioral process like mate selection. Brighter feathers demonstrate better nutritional condition in male House Finches, *Carpodacus mexicanus* (Hill *et al.*, 1994).

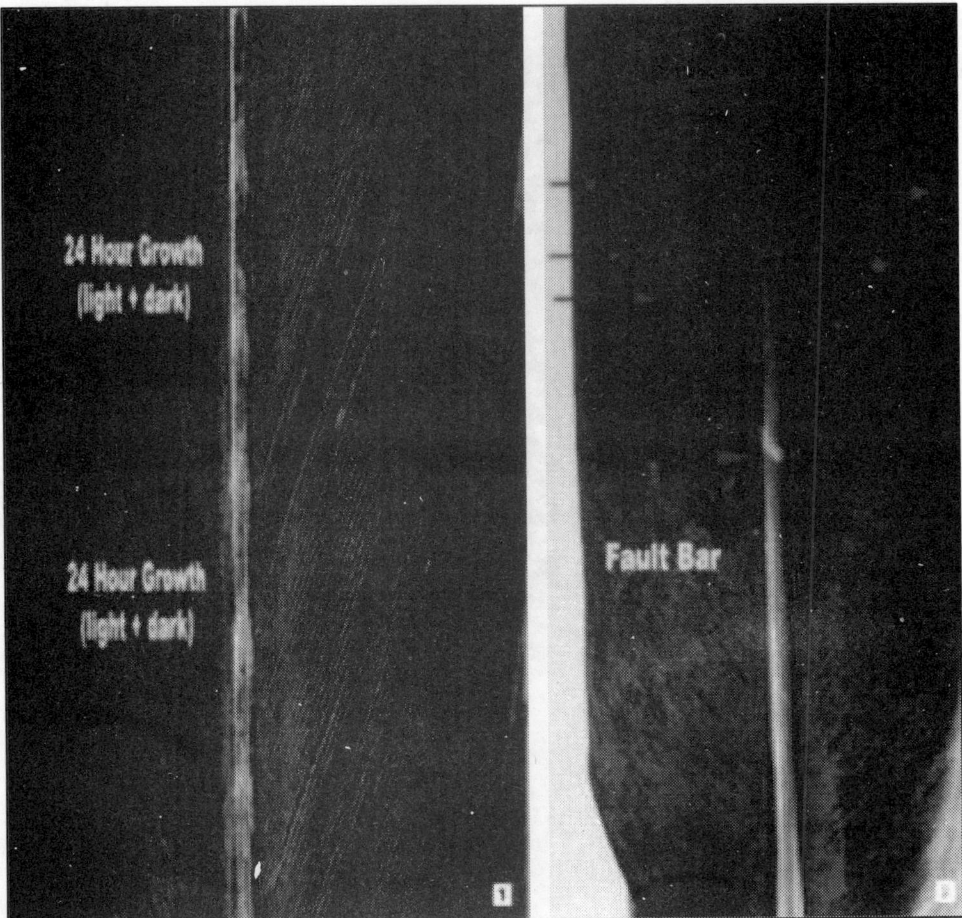

Plate 2: Growth and Fault Bars
Figure 1: Feather of Griffon vulture (*Gyps fulvus*) showing 24 hour growth bar including light and dark banks on rachis; Figure 2: Feather of Griffon vulture (*Gyps fulvus*) showing fault bars (Photos, *R. Yousef).*

Mate choice is influenced by the interaction of plumage coloration with the signaling environment along with the inherent properties of the sender (Heindl *et al.,* 2003).

The trait demonstrating male quality during sexual selection must be its honest signal to be an useful indicator (Grafen 1990). Structural plumage coloration is proved to be an honest indicator of male quality in Blue-black Grassquit, *Volatinia jacarina* (Doucet 2002). Siefferman *et al.* (2007) provided empirical evidence that environmental quality can influence the development of the structural coloration of feathers in Eastern bluebirds

Plate 3: Sample of Feather Growth.
Figure 1: Irregular; Figure 2: Optimal and Figure 3: Sub-optimal growth of feather influenced by feather condition in Graceful Warblers, *Prinia gracillis* (Yosef, 1997).

(*Sialia sialis*) and that males may be more sensitive to environmental fluctuations than females.

Several exogenous parameters like quality and quantity of food, ambient temperature and photoperiod were evaluated in order to elucidate

Figure 2.1: Different Coloured Feathers of Quail Species Inhabiting All Over the World.

Figure 1: Black francolin quail in Southern Asia; Figure 2: Blue breasted quail in China; Figure 3: California quail in North America; Figure 4: White Button Quail in America; Figure 5: Jungle bush quail in India, Sri Lanka, Pakistan and Nepal; Figure 6: Mountain quail in mountains of United States to Mexico.

1

Contd...

Figure 2.1–*Contd...*

3

Contd...

Figure 2.1–*Contd...*

their influence on the nutritional condition of the quail. This nutritional condition data was then correlated with the sexual selection trials in captive Japanese Quail. In paired trials, females showed their preference between a control and a treated/manipulated male.

A vast range of studies have been done explaining the factors responsible for the growth of feathers and fault bars. Ptilochronology has many facets; each of which opens a way to study a different aspect. We considered that since growth bars are the honest indicators of male quality (Takaki 2001), they might help females to select their mates qualitatively. Hence, we hypothesized that growth bar width, *i.e.*, the nutritional condition, would affect mate choice in Japanese Quail.

The Figure 2.1 shows the quail species from the different parts of the world with different coloured feathers. This shows that the climatic conditions of the surrounding environment including light, temperature, food available etc. might have an effect over the development and coloration of the feathers.

Thus, we decided to study such an interesting phenomenon in Japanese Quail.

Chapter 3

Japanese Quail: A Pilot Animal for Research

History of Japanese Quail

Coturnix were domesticated in Japan about the 11th century. They were initially raised as pet birds, but by 1900 they were widely used for meat and egg production. Japanese Quail were used as game birds around 1870 in the United States (Cain and Cawley 1914). They were also successfully introduced from Japan to America, Europe and the near and middle east between 1930s and the 1950s.

There are 45 species of quail worldwide and a total of 11 species in India. Most quail species are endangered worldwide. They are heavily hunted in their natural habitats for their vibrant feathers and meat. Hence, there is great importance in optimizing farming practices for human consumption in an effort to reduce the hunting pressure on the populations in the wild. Of these, the Japanese Quail (*Coturnix coturnix japonica*) is the domesticated quail commercially bred for human consumption.

Worldwide Geographical Distribution of Quail

Worldwide geographical distribution of quail is showing in Figure 3.1.

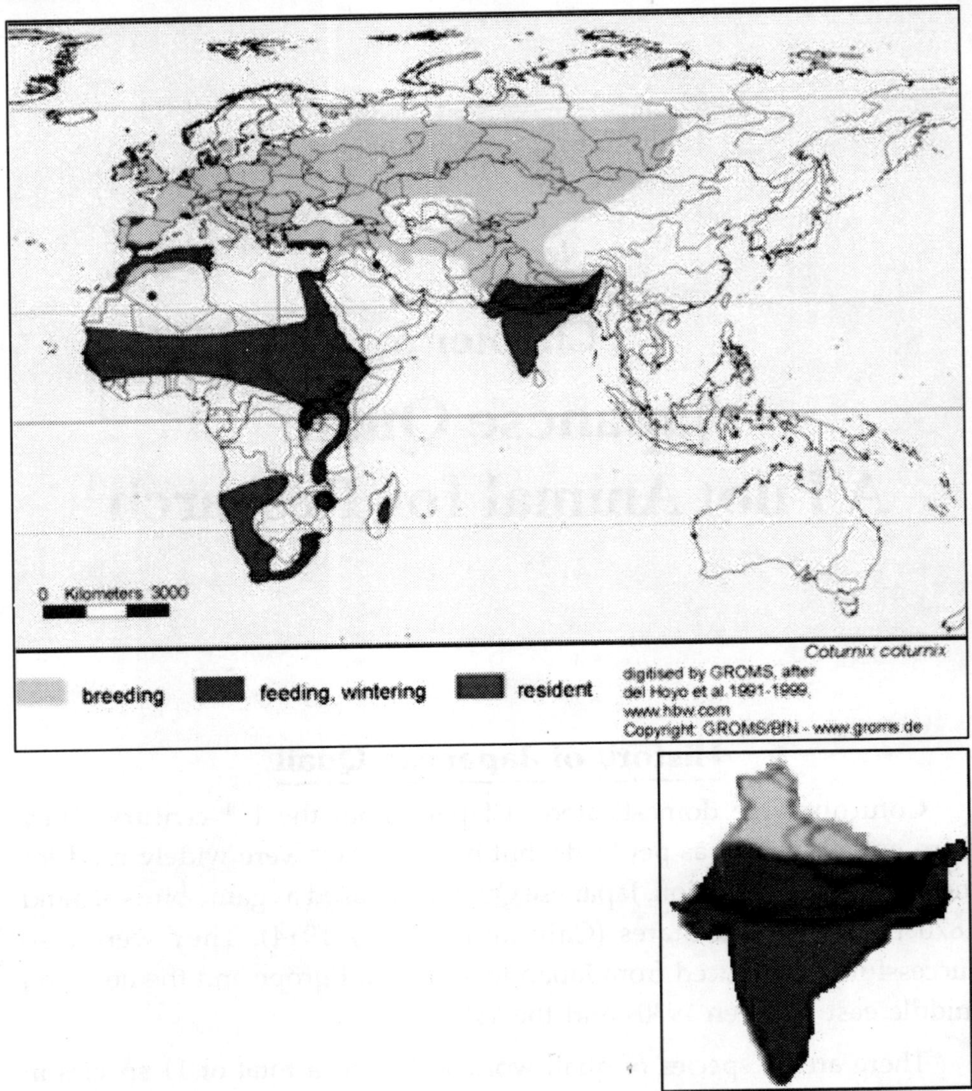

Figure 3.1: Geographical Distribution of Quail All Over the World.

Reasons to Choose Japanese Quail for Research

Extensive research on Japanese Quail carried out during 1900, showed that this was a valuable animal for avian research (Woodard *et al.*, 1973, Minvielle 2004). Japanese Quail is considered as a pilot animal for research purpose (Cain and Cawley 1914, Mills *et al.*, 1997) because it:

1. Is small in size, permits ease of housing and reduces feed costs,
2. Has a short lifecycle, thus physiological maturing and aging processes are rapid (Figure 3.2).
3. Has a sturdy nature, and is more disease resistant than other forms of poultry,
4. Respond favorably to laboratory manipulation,
5. Have higher metabolic rate than chicken and turkey, and
6. Respond rapidly to changes in the environment such as lighting or temperature fluctuations.

Disadvantages of the Species that are Compensated

1. They have high ammonia content in their droppings with an objectionable odour. This was reduced by proper ventilation and daily cleaning of cages.
2. A rapid decline in reproductive efficiency occurs after approximately 8 months of age. This was of least concern to the present study since we studied the behavioral responses of newly sexually matured quails.
3. Japanese Quail are more easily excited by sudden noises than other types of poultry. Extraneous influences were avoided by housing our experimental birds in a facility distant from human disturbances or environmental stochastic events.

Sexual Identification

1. Adult Male (Figure 3.3)

The adult male weighs about 100 to 140 grams. The males have a rusty brown throat and breast feathers. Males also have a cloacal gland, a bulbous structure located at the upper edge of the vent which secretes a white, foamy material. This is a unique gland and can be used to assess the reproductive fitness of the males (Cain and Cawley 1914, Ernst 1978). Woodard *et al*. (1973) had described adult male with a cinnamon-coloured feathers on the upper throat and lower breast region. It has a characteristic loud voice, described as "pick-per'awick" or "ko-turro-neex" (Sanford 1957).

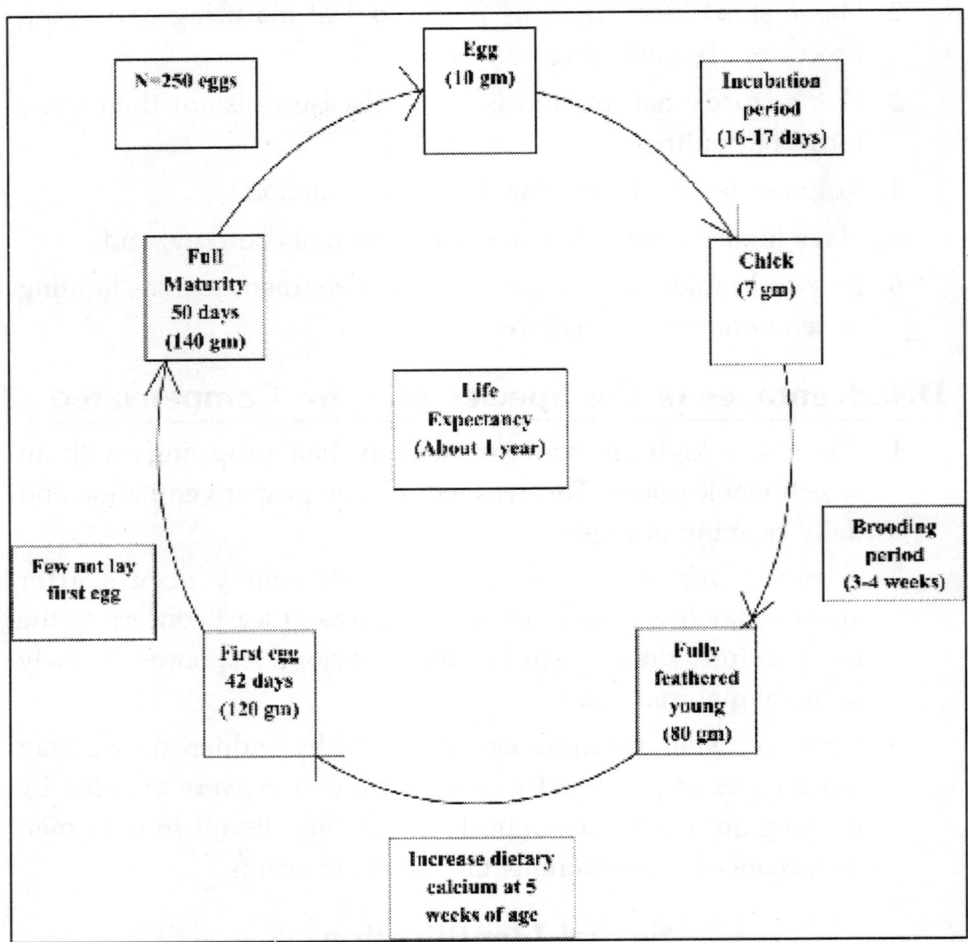

Figure 3.2: Schematic Representation of Life-cycle of Japanese Quail (Cain and Cawley 1914).

2. Adult Female (Figure 3.3)

Adult females are slightly heavier, weighing from 120 to 160 grams. The females are characterized by light tan feathers with black stippling on their throats and upper breasts (Cain and Cawley 1914, Ernst 1978). Woodard *et al*. (1973) had described females similar to the males in coloration except that the feathers on the throat and upper breast are long, pointed, and much lighter cinnamon.

Figure 3.3: Adult Male and Female Japanese Quail (*Coturnix coturnix japonica*).

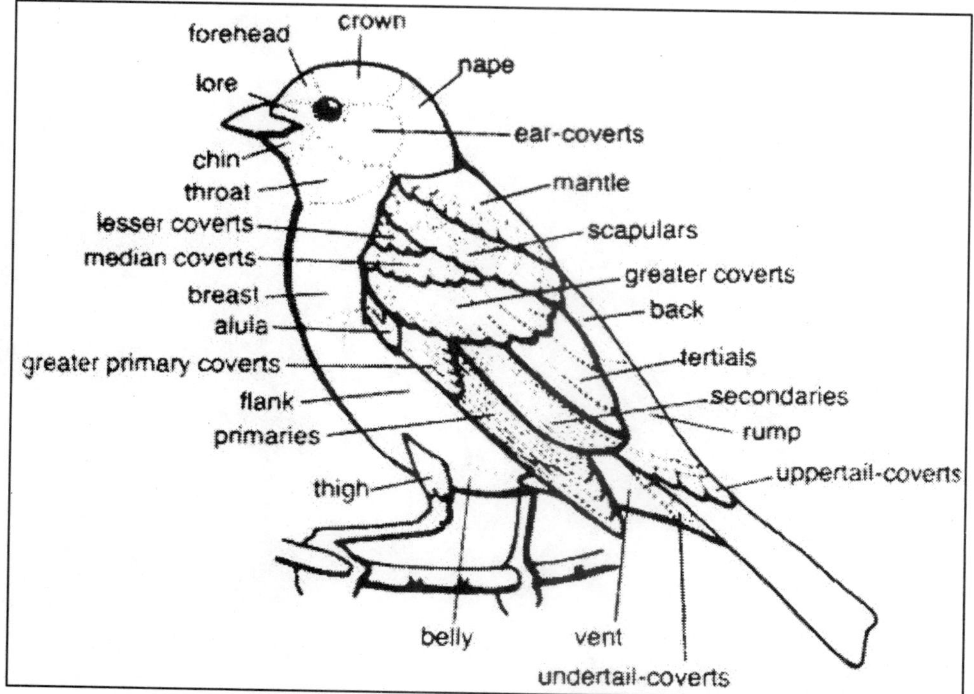

Figure 3.4: Diagrammatic Representation of a Bird with its Body Parts and Feathers.

Scientific Classification of Japanese Quail

- ☆ Kingdom - Animalia
- ☆ Phylum - Chordata
- ☆ Class - Aves
- ☆ Order - Galliformes
- ☆ Family - Phasianidae
- ☆ Genus - *Coturnix*
- ☆ Species - *coturnix*
- ☆ Subspecies - *japonica*

Chapter 4
Historical Review

1938-1990

Michener and Michener (1938) provided evidence that each pair of bands, that is each growth bar constitutes a 24-hour period of feather growth.

Wood (1950) suggested that in each growth bar darker bands are derived from material laid down in the follicle during the day hours and lighter bands consist of material laid down during the night hours.

Alantalo *et al.* (1984) checked if male and territorial factors influence female mate choice in Pied Flycatcher (*Ficedula hypoleuca*) in which arrival date explained more of female choice than male age, size and colour.

Andersson (1989) investigated sexual selection through female mate choice in the lekking Jackson's widowbird by applying multivariate selection analysis to observational data from four leks. It was found that the success of males in obtaining copulations was dependent on two components: display rate and lek attendance affected the number of female visits, whereas tail length primarily influenced the chance of copulating with a visiting female. Tail length was positively related to a measure of body condition, which was of interest with regard to the suggestions that sexual ornaments may serve as indicators of male viability.

Grubb (1989) provided evidence for the first time in Downy woodpeckers (*Picoides pubescens*), that feather growth bars indicate nutritional condition of free ranging birds concluding that males normally have a better nutritional condition as compared to females during winter.

Moller (1989) experimentally manipulated tail length in male swallow, *Hirundo rustica* and found that females prefer males with elongated tails over those with shortened tails. He further added that impaired foraging efficiency of tail-elongated males increased the frequency of fault bars in their tail feathers, probably as a result of food deficiency during moult.

Grubb *et al.* (1990) found that supplementary food significantly improved the nutritional condition of wintering birds of several bark-foraging species like Downy Woodpecker (*Picoides pubescens*), Carolina Chickadee (*Parus carolinensis*), Tufted Titmouse (*P. bicolor*), and White-breasted Nuthatch (*Sitta carolinensis*) by using daily growth of induced feathers as an index.

Hill (1990) done a field study which indicated that females House finch (*Carpodacus mexicanus*) prefer to mate with colourful males and that plumage brightness correlates with a males capacity for parental care and its genotypic quality.

Waite (1990) studied the effect of caching supplemental food on induced feather regeneration in wintering Gray Jays *Perisoreus canadensis*. He found that replacement feathers from jays with supplemental caches grew at a faster rate and were longer and heavier than those from jays that did not have access to supple-mental caches.

1991-2000

Grubb (1991) provided an experimental evidence for the fact that the width of growth bars on an induced feather is narrowed if a bird's diet is deficient during growth Carolina Chickadees (*Parus carolinensis*). It was found that feather growth, length and mass were significantly varied with age, sex and season. Also ambient temperature influenced feather growth more than photoperiod.

Grubb *et al.* (1991) studied the effect of seasonality, temperature, photoperiod, and bird's age and sex on daily growth of induced feather in Northern Cardinals (*Cardinalis cardinalis*). It was found that feather growth, length and mass were significantly varied with age, sex and season. Also ambient temperature influenced feather growth more than photoperiod.

Hill (1991) had done a field study which indicated that females in house finch (*Carpodacus mexicanus*) prefer to mate with colourful males and that plumage brightness correlates with male's capacity for parental care and perhaps its genotypic quality.

Machmer *et al.* (1991) documented the occurrence of fault bars in a population of nestling Ospreys *Pandion haliaetus* under natural condition. They found that fault bar formation declined linearly with age and increased symmetrically from outer to inner rectrices. Fault bar occurrence in relation to food shortage and handling effect was also examined but none of these had a significant effect on fault bar severity.

White and Kennedy (1992) demonstrated in American Tree Sparrow (*Spizella arborea*) that the duration of rectrix growth, measured in days, is controlled hormonally while the rate of growth, measured by growth bar width, is controlled nutritionally.

Yosef and Grubb (1992) used ptilochronology to show that a reduced number of hunting perches is correlated with reduced nutritional condition in Loggerhead Shrike (*Lanius ludovicianus*). Their results showed that density of hunting perches affects nutritional condition of non-breeding territorial shrikes.

Zuberbier and Grubb (1992) tested effect of cold temperature and/or wind on the growth of rectrices, independent of a bird's nutritional status in White- breasted Nuthatches. The results had shown no any significant effect of either wind, temperature or their interaction on the rate of feather regeneration.

Grubb and Yosef (1994) reported evidence that nutritional condition in Loggerhead Shrikes resident in south-central Florida is habitat specific.

They considered that pastures and citrus groves, respectively, are superior and marginal habitats for the species.

Hill and Montgomerie (1994) studied the interrelation between carotenoid based plumage color signals and nutritional condition in house finch (*Carpodacus mexicanus*). They compared the brightness of carotenoid-based plumage coloration with both feather growth rate and timing of moult. This study suggested that males growing brighter feathers are in better nutritional condition.

Ligon and Zwartjes (1995) showed that although females of red jungle fowl, *Gallus gallus* do discriminate between males, the plumage is not the target of the females' attention.

Yosef (1997) compared between habitats settled by Graceful Prinia (*Prinia gracilis*) and with the help of ptilochronology ranked their relative importance to avian populations. Their comparison showed that the reed beds were superior resource to the birds than the salt marsh.

Carrascal *et al.* (1998) studied the body condition and feather growth rate of Great Tits (*Parus major*) in relation to dominance in two contrasting Mediterranean localites during late autumn and early winter. He found that growth rate of induced feathers was slower in the locality that was colder and had less natural food availability, but was not clearly and directly related to dominance in both localities.

Grubb *et al.* (1998) assessed whether scrub-jay fledglings in different territories differed in nutritional condition, because differential nutritional condition could be responsible directly (starvation) or indirectly (depredation) for territory-specific mortality. These results suggest that nutritional condition plays a role in survival of young Florida Scrub-Jays and that relatively subtle differences in territory composition may contribute to variation in overall reproductive success.

Leonard and Zanette (1998) used paired choice tests to examine mate selection by female domestic chickens, *Gallus gallus domesticus*. Across trials, the wing-flapping rate of chosen males ranged widely (3–82 wingflaps/h) suggesting that females used a relative choice mechanism when selecting a mate.

Grether *et al.* (1999) showed that carotenoid availability in the wild limits the expression of sexual coloration in guppies (*Poecilia reticulata*), a species in which females prefer males with brighter orange carotenoid-containing spots.

Horak *et al.* (2000) investigated whether carotenoid-derived yellow plumage coloration of fledgling great tits (*Parus major*) depends on their growth conditions. Their experiments had shown that the quality and quantity of food during feather growth partly determine the expression of carotenoid-based plumage coloration.

2001-2011:

Hagelin and Ligon (2001) examined female mate choice in two congeneric species of New World quail, Gambel's quail, *Callipepla gambelii*, and scaled quail, *C. squamata*. They did not found male plumage to be the primary means by which female quail made mating decisions, instead females appeared to favour traits capable of signalling male condition, such as testosterone-mediated displays and body size.

Jenkins *et al.* (2001) tested the effectiveness of ptilochronology for assessing nutritional condition in male White-throated Sparrows (*Zonotrichiaa lbicollis*). They found that the birds fed an enriched diet weighed more, grew longer feathers, and had wider growth bars than birds receiving a diet lower in protein and calories. These results supported the validity of ptilochronology for directly detecting differences in nutritional status during feather regrowth, and point to a likely influence of social status, a density-dependent ecological stressor, on feather growth rates.

Pryke S., Andersson S. and Lawes M. (2001) demonstrated a strong female preference and net sexual selection for long tails, but marginal or no effects of color, morphology, or territory quality in red-collared widowbird (*Euplectes ardens*). Tail length explained 47 per cent of male reproductive success, an unusually strong fitness effect of natural ornament variation.

Takaki *et al.* (2001) examined for the first time the correlation between feather growth bar width and breeding parameters such as arrival date, pairing date and reproductive success in the Styan's Grasshopper warbler (*Locustella pleskei*). Their findings stated that the width of growth bars not only indicates an individual's body condition during moult, but also probably reflects its ability to maintain good body condition throughout the year.

Bortolotti *et al.* (2002) evaluated the frequency of fault bars in American kestrels stating that fault bars indicate an individual's susceptibility to stress, and are useful in predicting components of fitness.

Doucet (2002) used a principal components analysis (PCA) of four color characteristics to combine reflectance data into a single color score. The results had shown that structural plumage coloration is condition-dependent and are the honest signal of male quality in the Blue-black Grassquit, *Volatinia jacarina*.

Ballentine and Hill (2003) contradicted their own hypothesis that plumage intensity functions in female choice in Blue Grosbeaks (*Passerina caerulea*), proving that female Blue Grosbeaks chose males irrespective of plumage intensity. Thus, concluded that the intensity of blue coloration of male Blue Grosbeaks did not function as a criterion in female mate choice.

Calkins and Burley (2003) performed their experiments on California quail, *Callipepla californica*, establishing that female use male plumage characteristics when making mating decisions. Females choose on the basis of natural trait variation, but also responded to artificially manipulated traits.

Candolin (2003) had discussed in details about the different signals used by the females for choosing their mate. He also had described about the theoretical studies that had been done before to elaborate the evolution of multiple cues.

Heindl and Winkler (2003) tested that if the contrast between plumage color and its background affects mate choice in domesticated canaries (*Serinus canaria*) suggesting that females preferred the males which contrasted strongly against the background. They further added that not only inherent properties of the sender but also the interaction of bird color with the signaling environment influence mate choice.

Mahler *et al.* (2003) tested the association of carotenoid-based plumage color with both diet and sexual dichromatism. The patterns of carotenoid pigmentation in doves could not be attributed to sexual selection as determined using dichromatism.

Moreno *et al.* (2003) considered female characters of avian species for the sexual selection by males explaining the importance of egg shell pigmentation in sexual selection. According to his study, sexual selection influence avian egg colour by affecting male investment in species.

Senar *et al.* (2003) tested in the great tit (Parus major), a difference between carotenoid based structural coloration and melanin based coloration. The size of the melanin breast band was not correlated to nutritional condition as estimated by the rate of tail growth (ptilochronology), controlling for locality, age, sex, year and season effects. These results suggested that the expression of the two traits is regulated by different mechanisms.

Siefferman and Hill (2003) investigated in eastern bluebirds (*Sialia sialis*) whether male eastern bluebird ornamentation correlates with mate quality and reproductive success. The results suggested that plumage coloration is a reliable indicator of male mate quality and reproductive success. He had further added melanin-based and structural based plumages to be the honest signals of male quality and parental care that can be assessed by competitors or by potential mates.

Pendlebury *et al.* (2004) experimentally studied the effect of temperature variability on laying Japanese Quail at two levels of food supply and found that energy expenditure increased with higher daily temperature variation, and also after a sudden change in temperature. These findings demonstrated that variation in temperature affects energy expenditure with an effect on egg production. They also found food quality to affect energy expenditure, with birds fed low-quality food had higher energy expenditures than birds on high-quality food. They tried to explain further that feeding under low quality food treatment might require greater time and an increased processing in the gut required for a high-cellulose diet.

Siefferman and Hill (2005b) demonstrated that male eastern bluebirds (*Sialia sialis*) must trade-off conserving energy for production of future ornamentation versus expending energy for current reproduction.

Eraud *et al.* (2007) exposed male zebra finches (*Taeniopygia guttata*) to environmental stress by exposing them to two temperature regimes (6 and 26°C) over a 4 week period. Simultaneously, half of the males in each temperature group were supplemented with carotenoids, whereas the other half were not. The results suggested that ambient temperature contributed to the modulation of the expression of carotenoid-based ornaments.

McGlothin *et al.* (2007) examined the effects of diet quality on the size and brightness of the tail-white patch in male dark-eyed juncos (Junco hyemalis). The results suggested that variation in diet quality affected an attractive white plumage pattern which may be stronger in stressful environments.

Siefferman and Hill (2007) provided empirical evidence that environmental quality including parental care, brood size, nestling condition and seasonality can influence the development of the blue structural coloration of feathers in eastern bluebirds (*Sialia sialis*) and males are more sensitive to environmental fluctuations than females.

Freeman-Gallant *et al.* (2009) explored sexual selection and age- and condition-dependent signaling in the common yellowthroat (*Geothlypis trichas*), a sexually dichromatic warbler with two prominent plumage ornaments—a melanin-based, black facial "mask" and carotenoid-based, UV-yellow "bib."

Holveck *et al.* (2009) provided the first example where an experimental manipulation of female quality induced a mating preference for low-quality males. Zebra finches (Taeniopygia guttata) reared in small or large experimental broods became high- or low-quality adults, respectively. The high-quality females preferred high-quality males' mate-advertising songs, while all low-quality females preferred low-quality males' song.

Sundev *et al.* (2009) evaluated the nutritional condition of nestling Upland Buzzards (*Buteo hemilasius*) by using the non-invasive ptilochronology technique. They found that the growth bar width of nestlings was significantly different and wider in vole-rich areas than in areas lacking voles.

Visser *et al.* (2009) tested the direct effect of temperature on laying dates in great tits (*Parus major*). They found that temperature has a direct effect on timing of breeding, an important step towards assessing the implication of climate change on seasonal timing.

Matysiokov*a et al.* (2010) tested the usefulness of ptilochronology in the study of melanin- and carotenoid- based ornaments in the Great Tit (*Parus major*). They found no relationship of growth bar width to the size of the melanin ornament but a negative relationship to the chroma of the carotenoid ornament. Thus, the growth rate of tail feathers declined as carotenoid levels in breast feathers increased.

Chapter 5
Application of Ptilochronology

Forty-five females and 40 males approaching maturity (3-4 weeks) were obtained from Chandrapur Quaillary, India. The males and females were housed separately in groups of 10-12 for acclimatization in a well ventilated room (20ft. x 20ft), at Sevadal Research Centre, Nagpur (21.07° N, 79.27°E), located centrally in India. During the acclimatization process (Plate 11 - figure 1), the birds had access to ad libitum control feed (Table 5.1) with balanced composition and water, light: dark (L: D) 15: 9 and temperature approximately 30°C (±5°C); for 8 days (Recommended by Central Avian Research Institute (CARI), Izatnagar, India). We measured body mass, body temperature, length of tail feather and daily consumption of feed of the birds (Plate 10; Table 5.1, 5.2).

Housing

For experimentation, the males were caged individually in specially designed cages (1 x 1.5 x 2 ft, (Figure 5.2(A); Plate 4 - Figure 3), arranged in two rows, such that control (C) and experimental (E) birds faced each other. Each cage was partitioned with removable aluminium sheets to form five compartments for individual accommodation of experimental birds. These cages composed of four separate sides (Figure 5.2 (C); Plate 5 - figure 2).

CAGE ARRANGEMENT

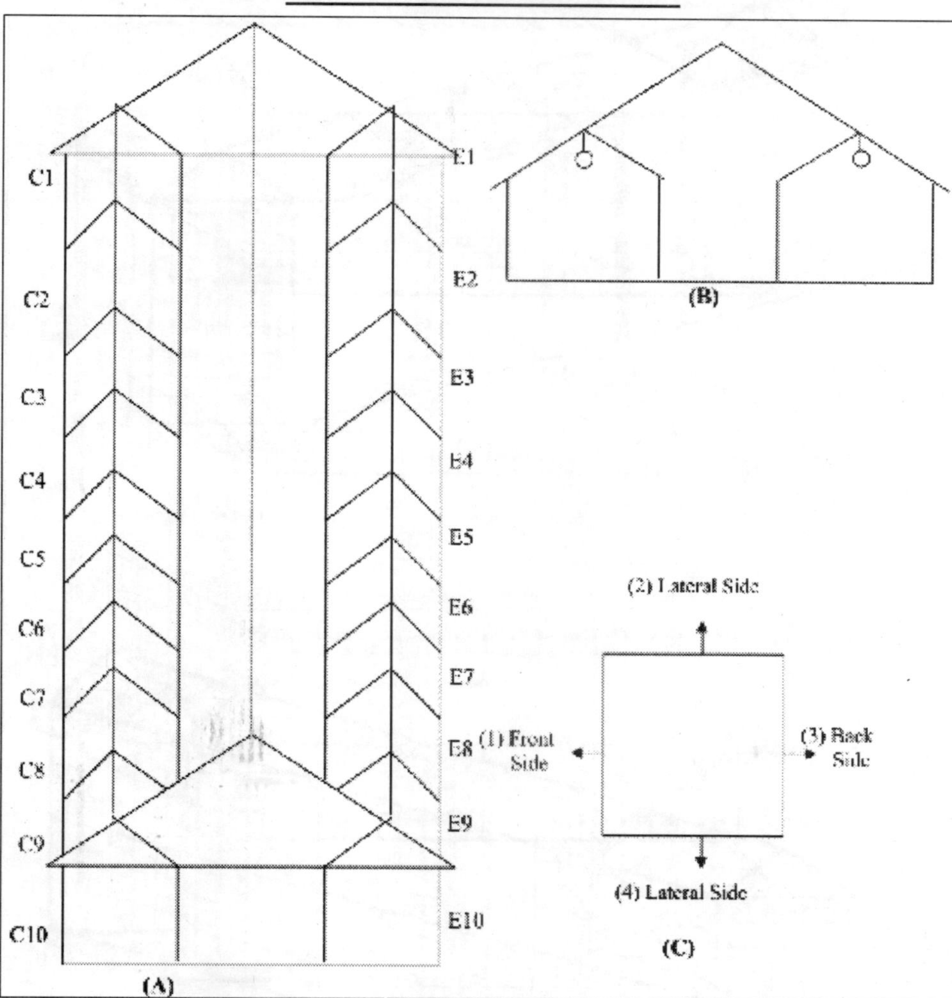

Figure 5.1: (A) Experimental arrangement of cages; (B) Side view of 2 cages placed in front of each other showing light arrangement; (C) Top view of 1 cage showing each side of cage designed specially for some purpose–1) Front side-transparent glass; 2) and 4) two lateral sides-opaque; 3) Back side-door.

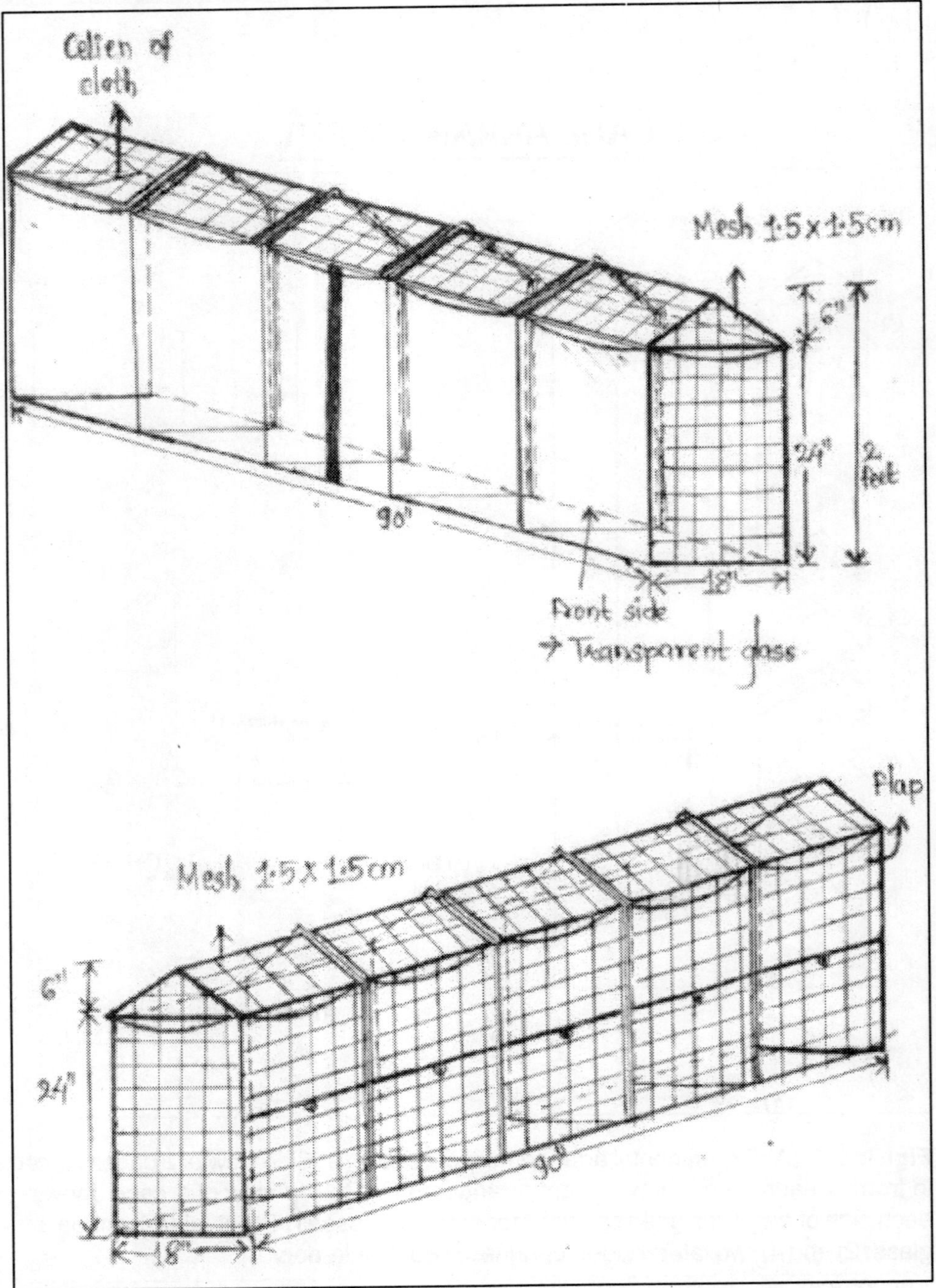

Figure 5.2: Schematic Representation of Cage with Measurements Designed for Individual Housing of Males for Experiments.

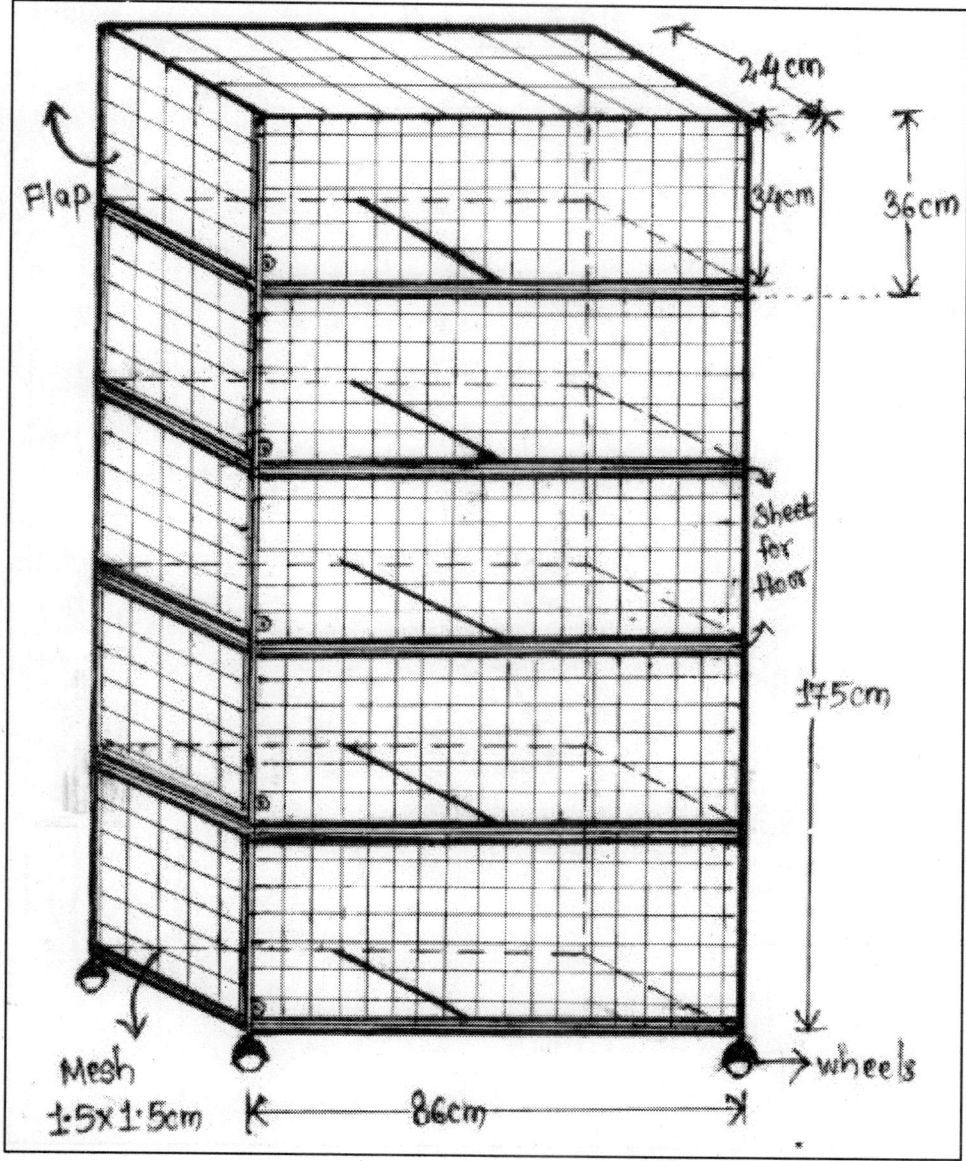

Figure 5.3: Schematic Representation of Vertical Cage with Measurements Designed for Accommodation of Females within Less Space.

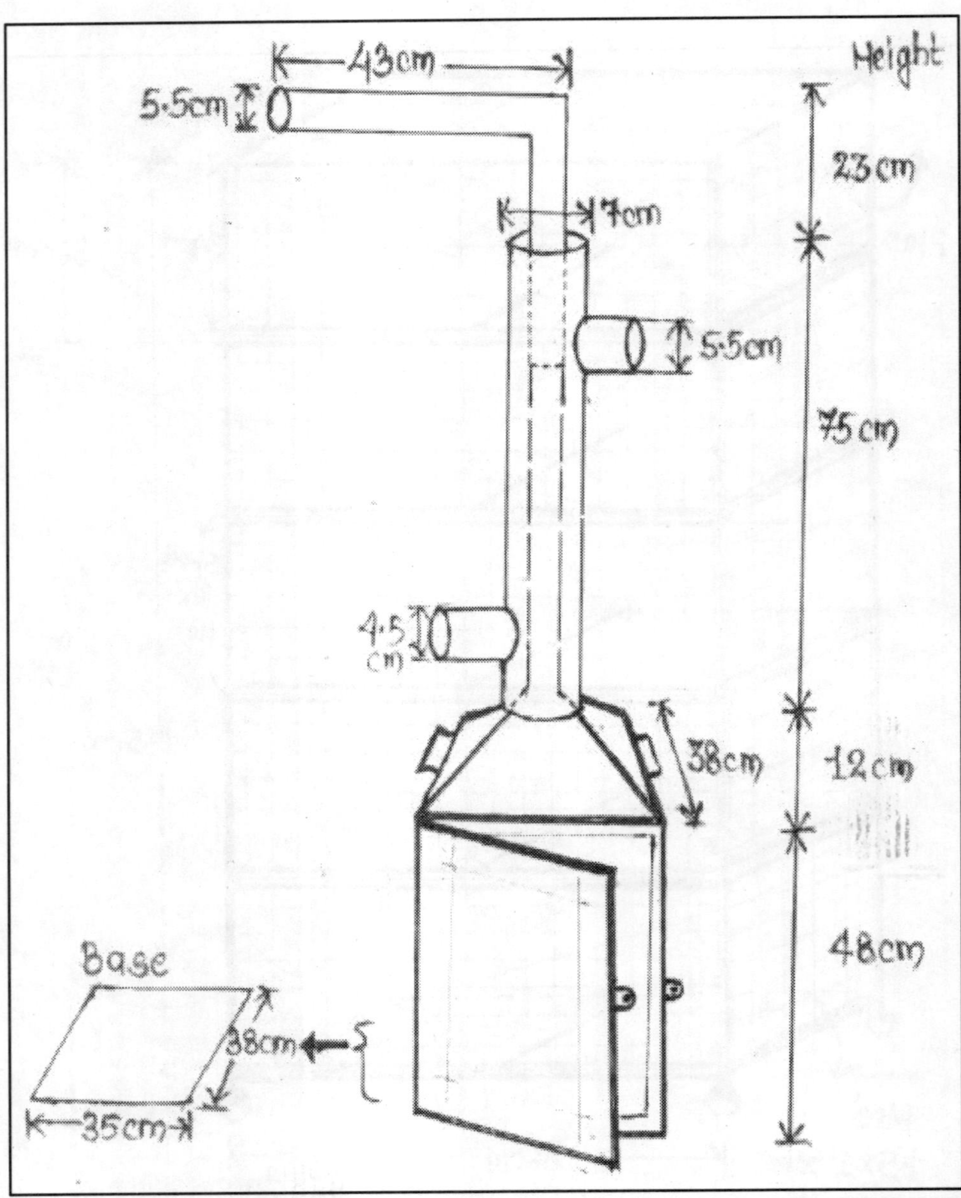

Figure 5.4: Schematic Representation of Equipment with Measurements Designed for Regulating Temperature Conditions during Fourth Experiment of Ambient Temperature.

a. Front (facing opposite cage) is of transparent glass for direct observation of each other (*i.e.*, male and female) during the selection experiment. During the non-trial periods it was covered with a curtain.

b & c. The two lateral sides (2nd and 4th) are opaque for separation and isolation purpose of males from all others.

d. The back side has a door with a net, for maintenance purposes.

The central passage (7.5 x 1.5 x 2 ft), created by placing the cages in two rows (3 above), was used only by females during the selection experiments.

Feed was prepared for the control and experimental males as well as females by mixing the ingredients (Plate 9 - figure 1; Table 5.1).

Table 5.1: Composition of 1 kg of Feed as per Requirements of the Experiment.

	Feed Ingredients	*Concentration (in gms)*
CONTROL	Concentrate	500
	Wheat Bran	100
	Maize	300
	Broken Rice	100
18 per cent PROTEIN DIET	Concentrate	400
	Wheat Bran	200
	Maize	250
	Broken Rice	150
16 per cent PROTEIN DIET	Concentrate	300
	Wheat Bran	250
	Maize	250
	Broken Rice	200
LAYER	Concentrate	350
	Wheat Bran	50
	Maize	340
	Calcium Powder	60
	Broken Rice	200

Table 5.2: Drinking Water Supplements.

Sl.No.	*Water Supplements*	*Quantity Used/ 100 Birds*	*Nutritional Value*	
1.	Stresvel	10 ml	**Nutritional value/ml**	
			Vitamin A	50,000 I.U
			Vitamin D_3	5,000 I.U
			Vitamin E	30 I.U
			Vitamin C	100 mg
			Carbohydrate, Organic nutritive base	Q.S.
2.	Ventriplex-M	10 ml	**Nutritional value/5 ml**	
			Vitamin B_2	1.25 mg
			Vitamin B_6	0.62 mg
			Vitamin B_{12}	6.25 mcg
			D-Panthenol	1.25 mg
			Niacin	37.50 mg
			DL-Methionine	5.00 mg
			Aqeous base	Q.S.
3.	Venlyte powder	10 gm	**Composition per cent W/W**	
			Sodium chloride	1.0
			Calcium lactate	1.1
			Magnesium sulphate	0.9
			Potassium chloride	3.0
			Sodium bicarbonate	1.0
			Sodium citrate	2.5
			Vitamin C	1.0
			Dextrose	78.2
			Carrier	Q.S.
4.	Glucon-D	10 gm	**Nutritional value/100 gm**	
			Calcium	170 mg
			Phosphorus	100 mg
			Vitamin D	300 I.U.
			Energy value	360 k cal
			Protein	0 gm
			Carbohydrates	90 gm
			Fat and all types of fatty acids	0 gm
			Cholesterol	0 gm

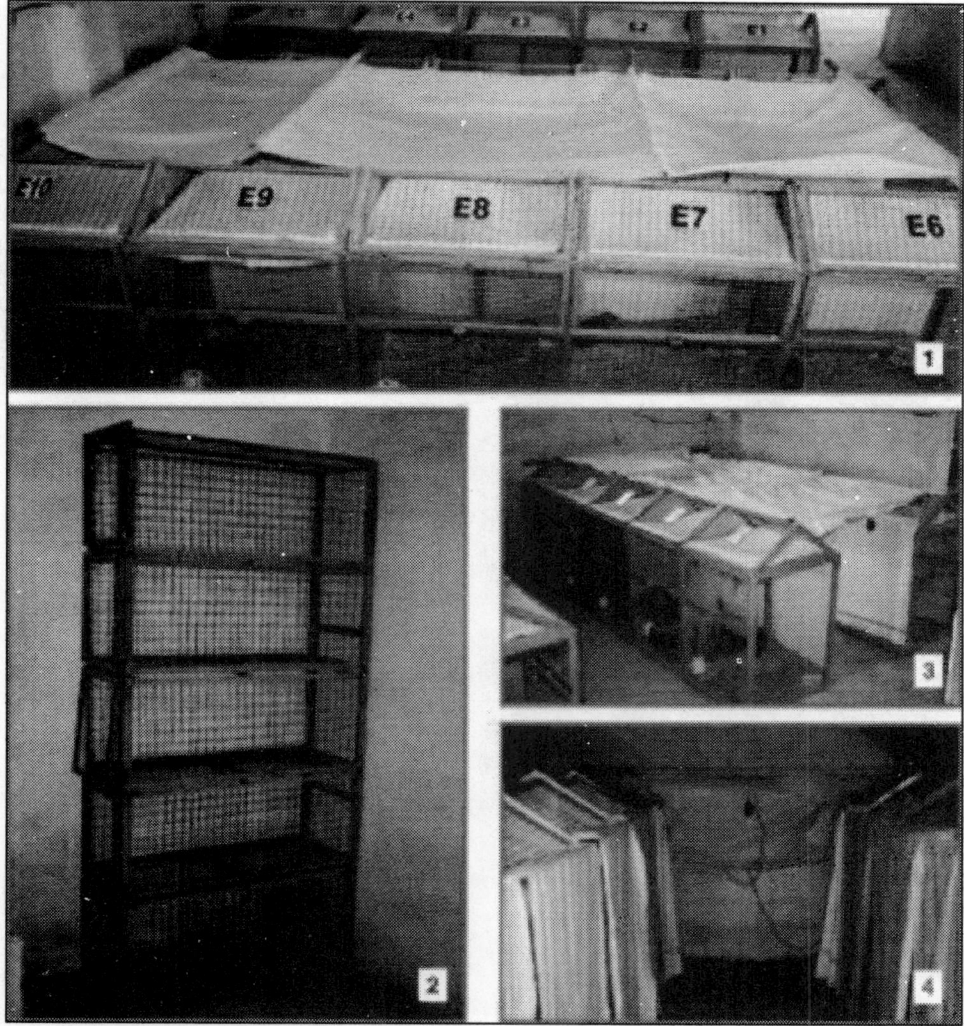

Plate 4: Experimental Setup

Figure 1: Experimental arrangement showing single male Japanese Quail in separate compartments of the cage; Figure 2: Five storeyed vertical cage designed for accommodation of female Japanese Quail; Figure 3: Cage arrangement for experiments.

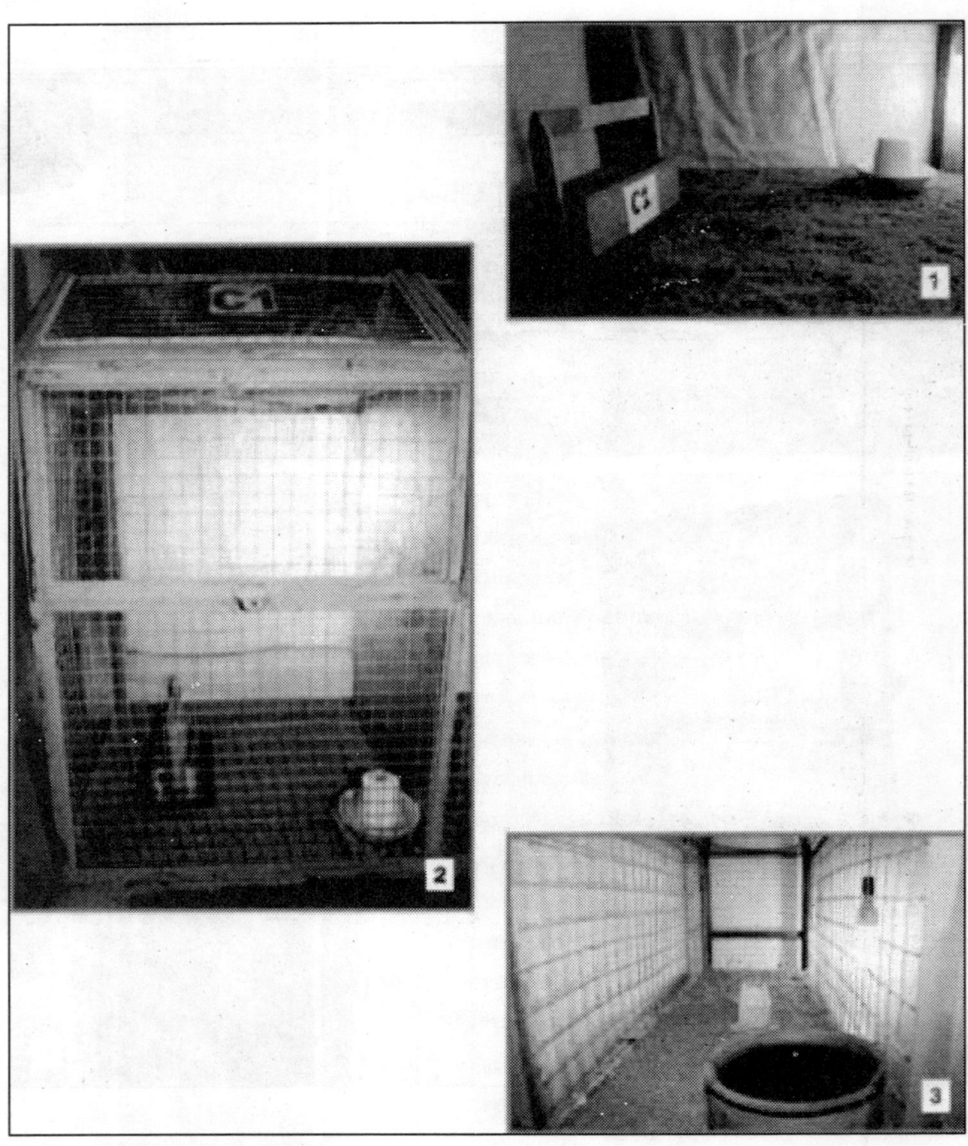

Plate 5: Cage Arrangement.

Figure 1: Floor in the edge prepared with mix of sand and wood husk; Figure 2: Cage compartment for individual housing of experimental experimentation; Figure 3: Cage compartment for females before procurement of the birds.

Plate 6: Other Necessities.
Figure 1: Cages for external transportation; Figure 2: Omnijet inverter to avoid fluctuations in electric supply; Figure 3: Cages for internal transportation.

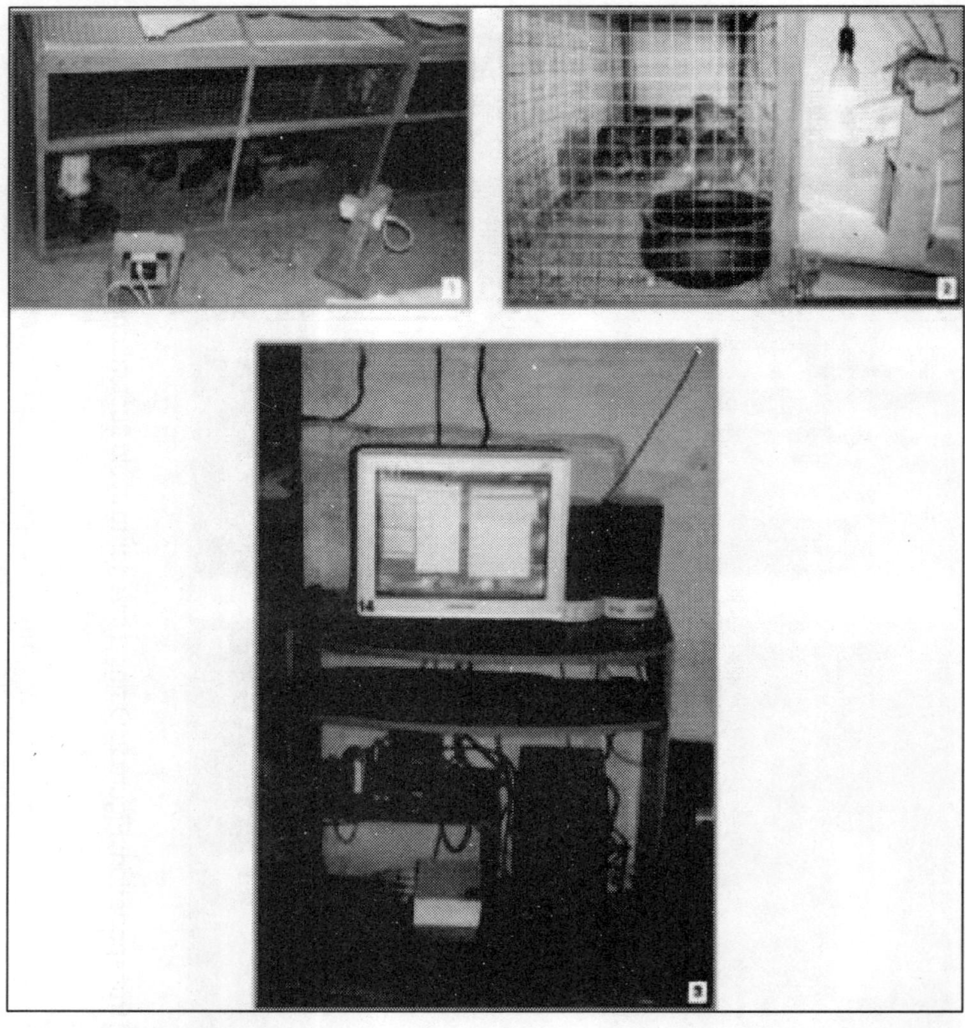

Plate 7: Closed Circuit Television Camera.

Figure 1: Closed circuit television (CCTV) cameras with audio arranged to record the moments of the male Japanese Quails without any disturbance; Figure 2: Arrangement of closed circuit television (CCTV) cameras in the cage compartment of female Japanese Quails; FIgure 3: Life display of camera recordins on computer and storage devices.

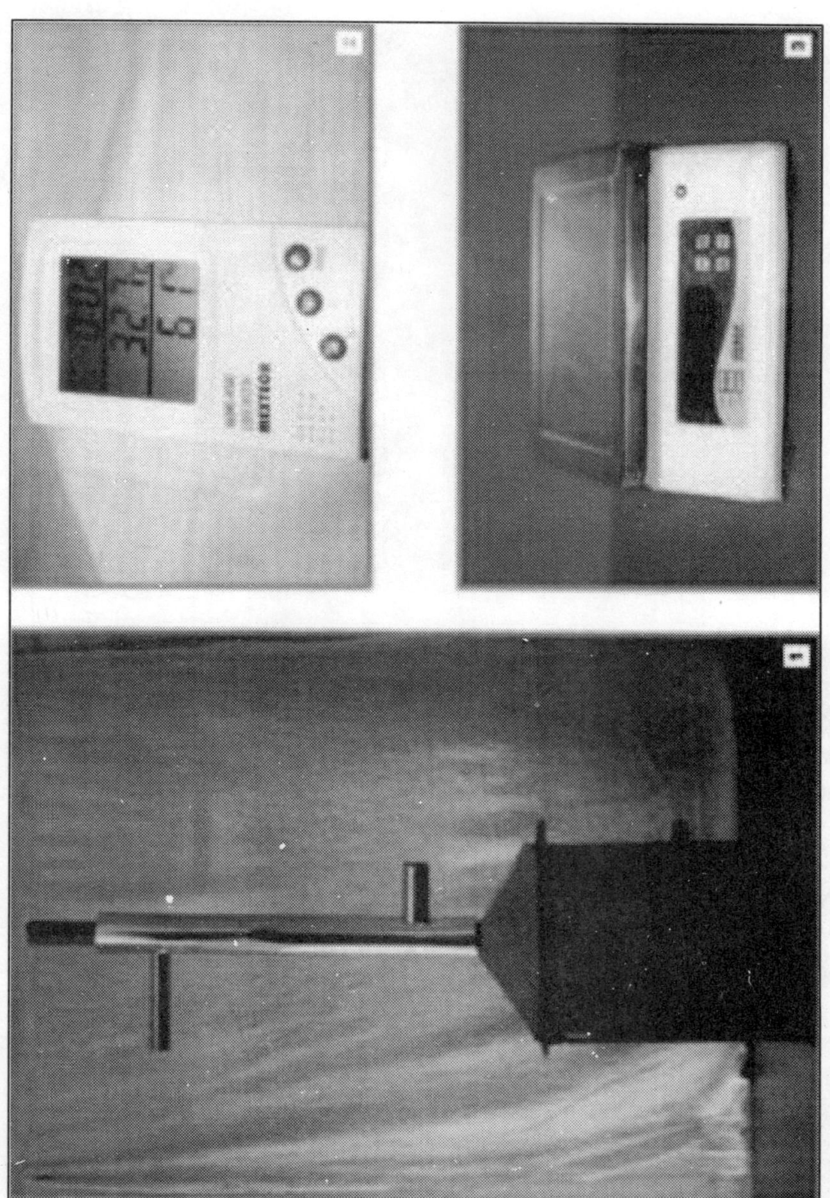

Plate 8: Equipments Used.
Figure 1: Equipment designed for temperature regulation; Figure 2: Digital thermo-hygrometer; Figure 3: Digital one pan weighing balance.

Plate 9: Feed and Water Supplements.
Figure 1: Ingredients used to prepare feed on quails; Figure 2: Wooden feeder designed to reduce feed loss; Figure 3: Water supplements added in the drinking water for supplementation of vitamins and electrolytes to the birds: Figure 4: Drinker for individual experimental bird.

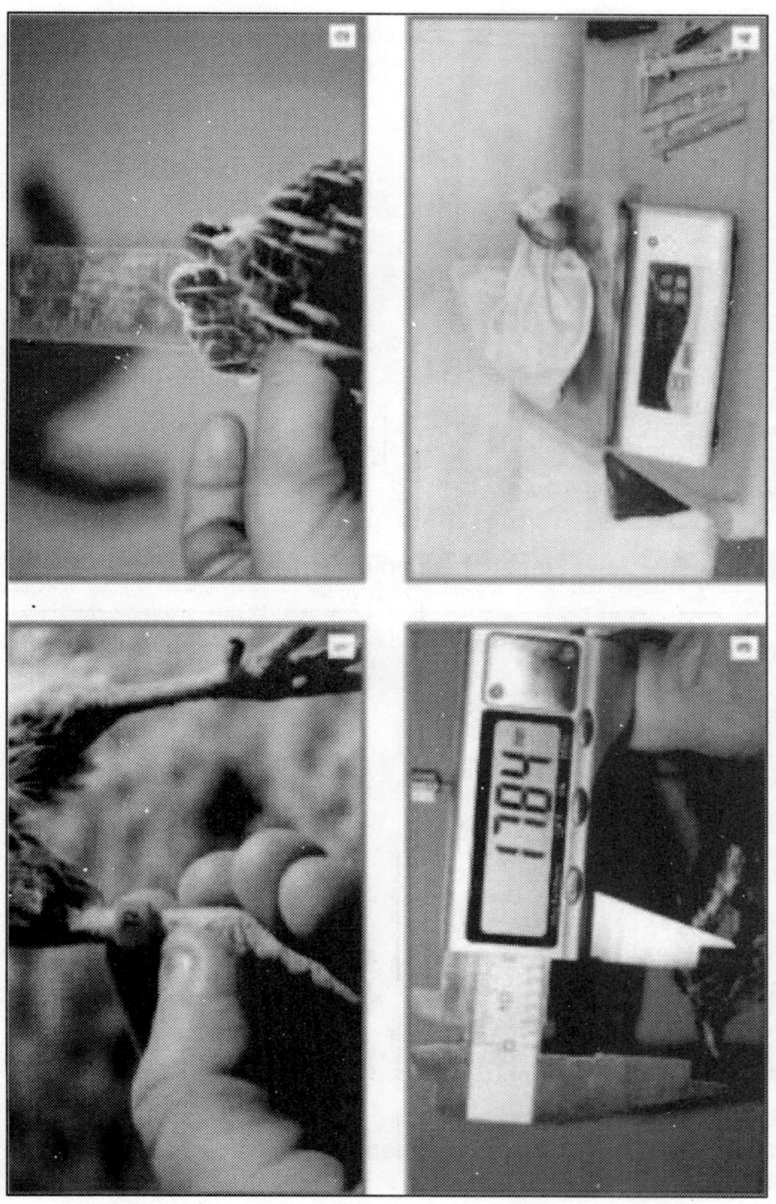

Plate 10: Ringing and Measurements.
Figure 1: Experimental bird showing labeled rings for individual identification; Figure 2: Measurement of length of central rectrices (tail feathers); Figure 3: Beak length measured by electronic Vernier caliper; Figure 4: Body mass measured with the help of electronic one pan weighing balance.

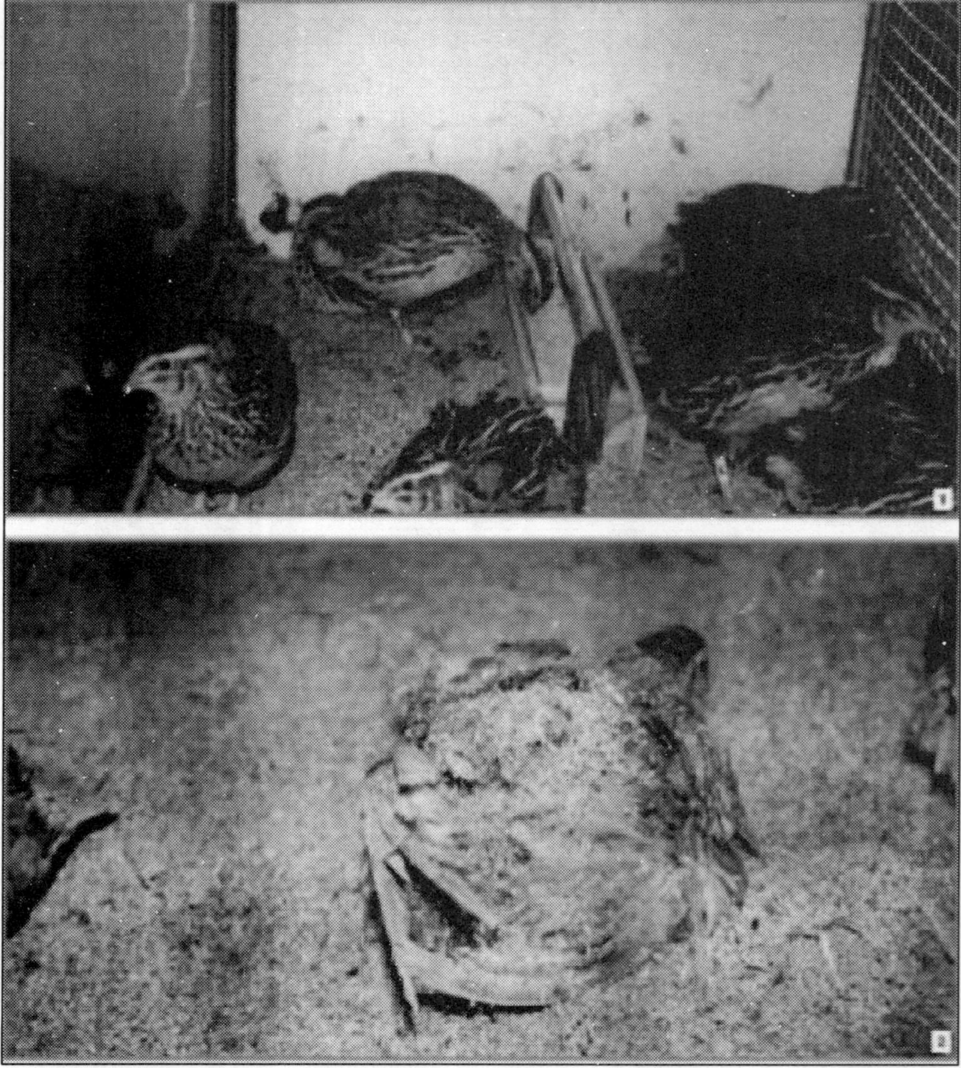

Plate 11: Acclimatisation.
Figure 1: Acclimatisation of subjects before experiments; Figure 2: Birds taking sandbath during acclimatisation.

Water supplements were added to the drinking water as under (Plate 7 - Figure 3; Table 5.2).

Procedure

The experiment was conducted in 4 different, monitored parameters:

- ✰ Food quality
- ✰ Food quantity
- ✰ Photoperiod
- ✰ Ambient temperature.

The sequence of the experimental procedure was:

1. Birds were acclimatized to experimental surroundings and apparatus
2. Then control rectrices plucked
3. Birds were held under different regimes (diet and photoperiod)
4. Then selection experiments conducted
5. Finally, induced feathers plucked for ptilochronology evaluation.

5.1 Experiment 1–FOOD QUALITY (Plate 12)

Control and the experimental males were fed with different quality food. The controls were given a balanced diet whereas the experimental were given starter diet of higher quality. The daily quantity given to all the males was 30gms, since their daily feed consumption was 20gms (±3 gms). The photoperiod for all birds was also same, L: D=15:9. Also the ambient temperature was maintained at 30°C (±3°C). Under these conditions, the feathers of the males (especially the central rectrix) were allowed to grow for 15-18 days, and when fully grown; selection experiments were undertaken.

A female was placed in the passage of the experimental area and exposed simultaneously to a control and experimental male. The female was allowed to remain 10 to 20 min with each of the males. The time allocated was on the basis of the time the female spent with each/either of the males. The greater the amount of time spent by the female with a male, greater was the chances of mating between them. This allowed us to interpret her selection of a particular male for mating (Plate 12 - figure

Plate 12: Experiment 1–Food Quality.
Figure 1: Different quality food prepared to feed controls (Control) and experimental (Starter) birds in first experimental of food quality; Figure 2: Selection experiments performed after completion of first experiment.

2). In a similar fashion, sequentially six females were introduced to the same pair of males. This was repeated for each of the ten pairs of males, *i.e.*, a total of 60 sexual selection experiments. These observations were recorded by using closed circuit television (CCTV) camera (Plate 7, 17).

After the completion of the sexual selection experiments, the fully grown central rectrices of the control and experimental males were simultaneously plucked. These feathers were stored in separate, individually marked, zip-lock plastic bags (2"x3") in deep freezer to prevent feather damage by bacterial or parasitic growth. Ptilochronology was done by holding the feather from both of its ends on a background of black color, below which a card is kept for documenting and indexing the growth bars. With the help of a small, thin mounting pin we pierce through the feather the underlying card. Initially, the length of the feather is measured by marking both the tips of the feather; then we pierce ten consecutive dark-and-light bands from the distal edge of the feather. By taking an average of the ten readings, an average growth bar width is calculated with the help of Electronic Digital Caliper (0-150 mm/0-6"; ±0.01mm/ 0.001"). Thus, the average growth bar width for all the males was calculated. The fault bars in all the feathers were also identified and counted (Plate 13).

5.2 Experiment 2–PHOTOPERIOD (Plate 14)

The second set of experiments included a variation in the light period to which the birds were exposed. These experiments were done in two parts;

PART-I: With a photoperiod of L: D=17:7, and

PART-II: With photoperiod L: D=13:11.

The balanced control diet was given to the birds during this experiment in a fixed quantity of 30 gm per quail per day. The temperature was maintained at 30°C (±3°C).

5.3 Experiment 3–FOOD QUANTITY (Plate 15)

The only variable in this experiment was food quantity. Since the effect of the parameter was to be studied through ptilochronology, the variation was in the quantity of protein content in the feed. This experiment was also done in two parts;

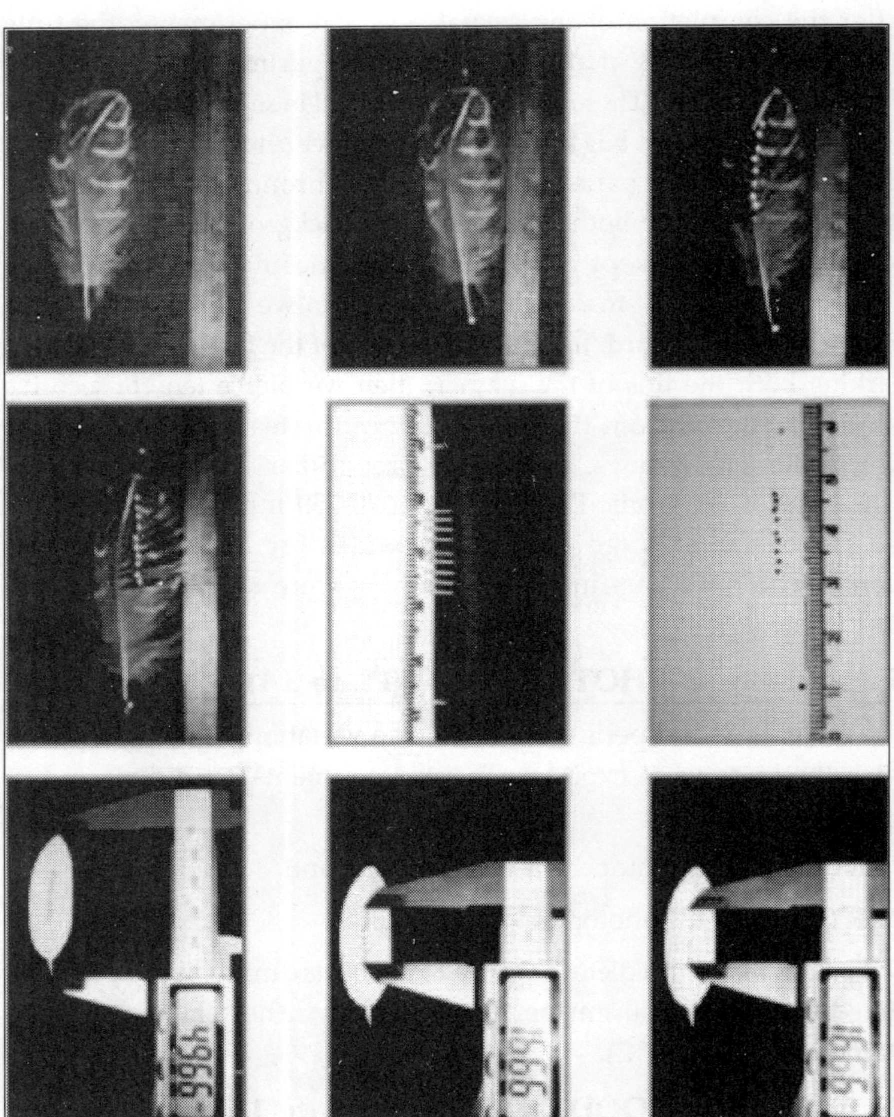

Plate 13: Ptilochronology.
Figures: Methodology for the measurement of growth bar width of a feather.

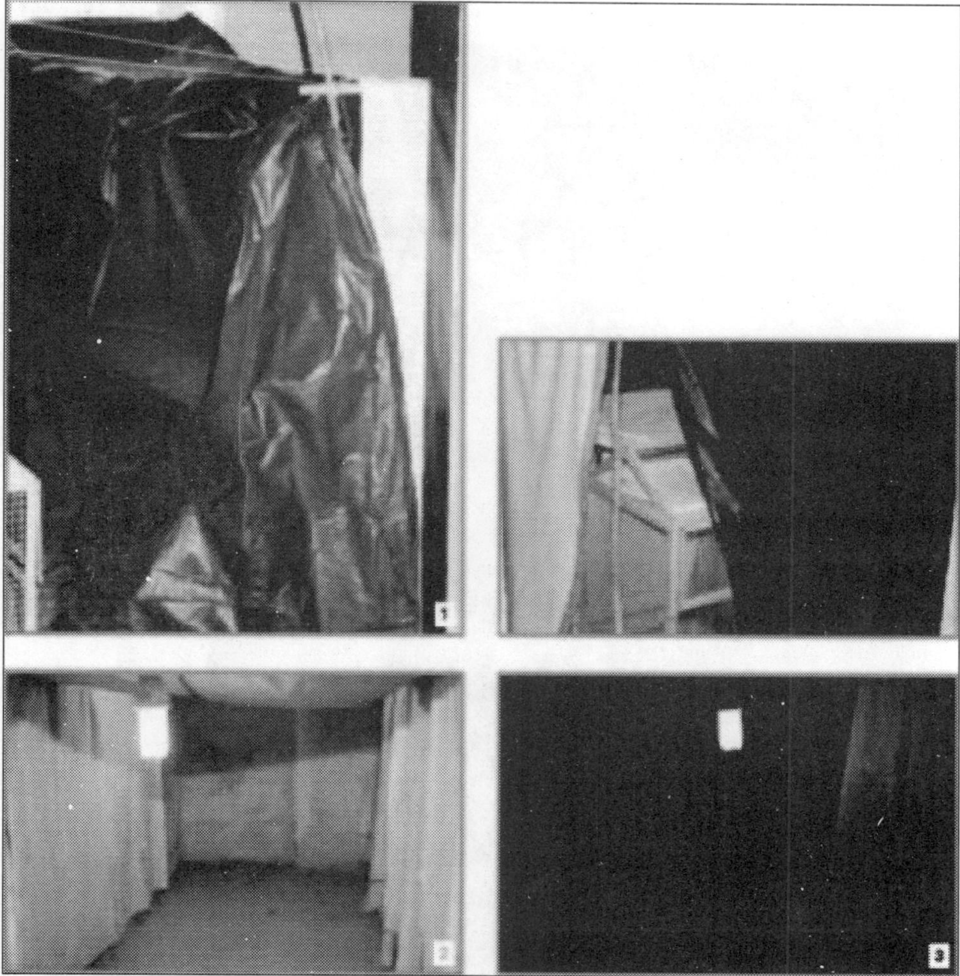

Plate 14: Experiment 2–Photoperiod.
Figure 1: Experimental arrangement for creating complete darkness during photoperiod experiment; Figure 2: Light hours in the photoperiod experiment; Figure 3: Darkness maintained during the dark hours of the experimentation.

Plate 15: Experiment 3–Food Quantity.
Figures: Varied protein quantity feed prepared for food quantity experiment–18 per cent protein diet for controls; 16 per cent protein diet for experimental males.

PART-I: controls were fed a normal diet, and manipulated quail were fed with 18 per cent protein diet.

PART-II: controls were fed a normal diet, and manipulated quail were fed 16 per cent protein diet.

All other parameters were the same, as detailed above.

5.4 Experiment 4–AMBIENT TEMPERATURE (Plate 16)

In this experiment, we manipulated the ambient temperature. An ambient temperature of 30°C (±2°C) was maintained for the control males but was raised to 35°C (±2°C) for the experimental males (Figure 3.5). After the full growth of the feathers, selection trials were conducted and feathers were plucked for ptilochronology.

The sexual selection decisions of the females were noted, for each of the four experiments. They were further analyzed to study the impact of the varying environmental parameters on the sexual selection of the female quail. Because the sample size was small ($n<30$), we used 'Student's t-test'. Also, the average daily growth bar width of the plucked rectrices of control and treated males were more effectively interpretated by using a 'paired t-test'. We accepted $P>0.05$ as the minimum level of significance.

For this experiment we designed a special instrument to regulate temperature (Figure 3.4). The details are as below,

Principle and Working

The basic principle behind the instrument is regulation of temperature by using charcoal. The charcoal is burned in the furnace (35 cm x 38 cm base) like pot in the middle, with small pores on the door to receive oxygen for burning. The heat will be accumulated since no any outlet is there. The smoke from the burning will be removed with an aluminium pipe (98cm long) connected to the upper part of the furnace. This pipe is again covered with another broad steel pipe (75 cm long and 7 cm wide) with inlet (4.5 cm wide) and outlet (5.5 cm wide). The inlet allows the surrounding air to come inside and then it travel side by side surrounding the middle pipe of smoke and heat. This makes the air hot which is then released to the desired area or room from the outlet connected with the aluminium pipes.

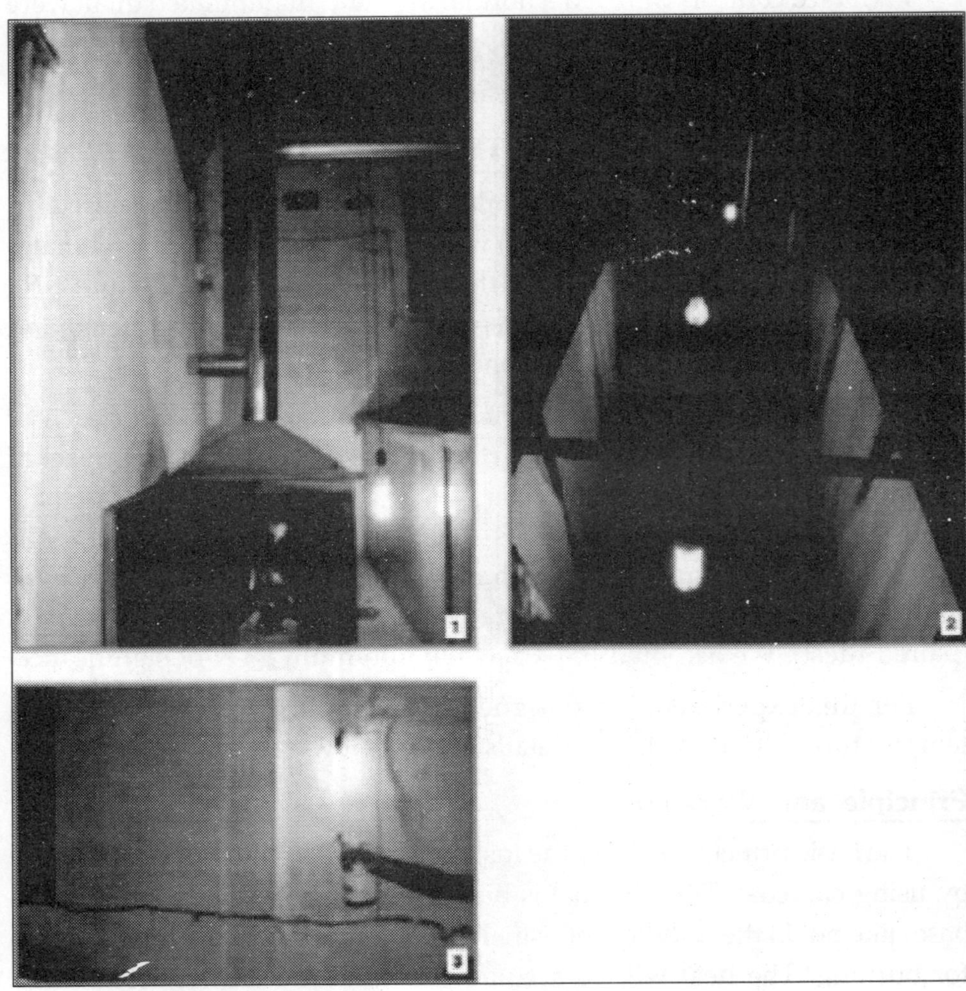

Plate 16: Experiment 4–Ambient Temperature.
Figure 1: Equipment designed to regulate constant temperature conditions; Figure 2: Vertical pipe releasing hot air to maintain temperature in the cage compartment of experimental birds: Figure 3: Spraying of water done for maintenance of surrounding temperature.

Benefits

1. The instrument does not work on electricity.
2. It is very efficient and controlled method of temperature regulation.
3. It is cheaper to install.

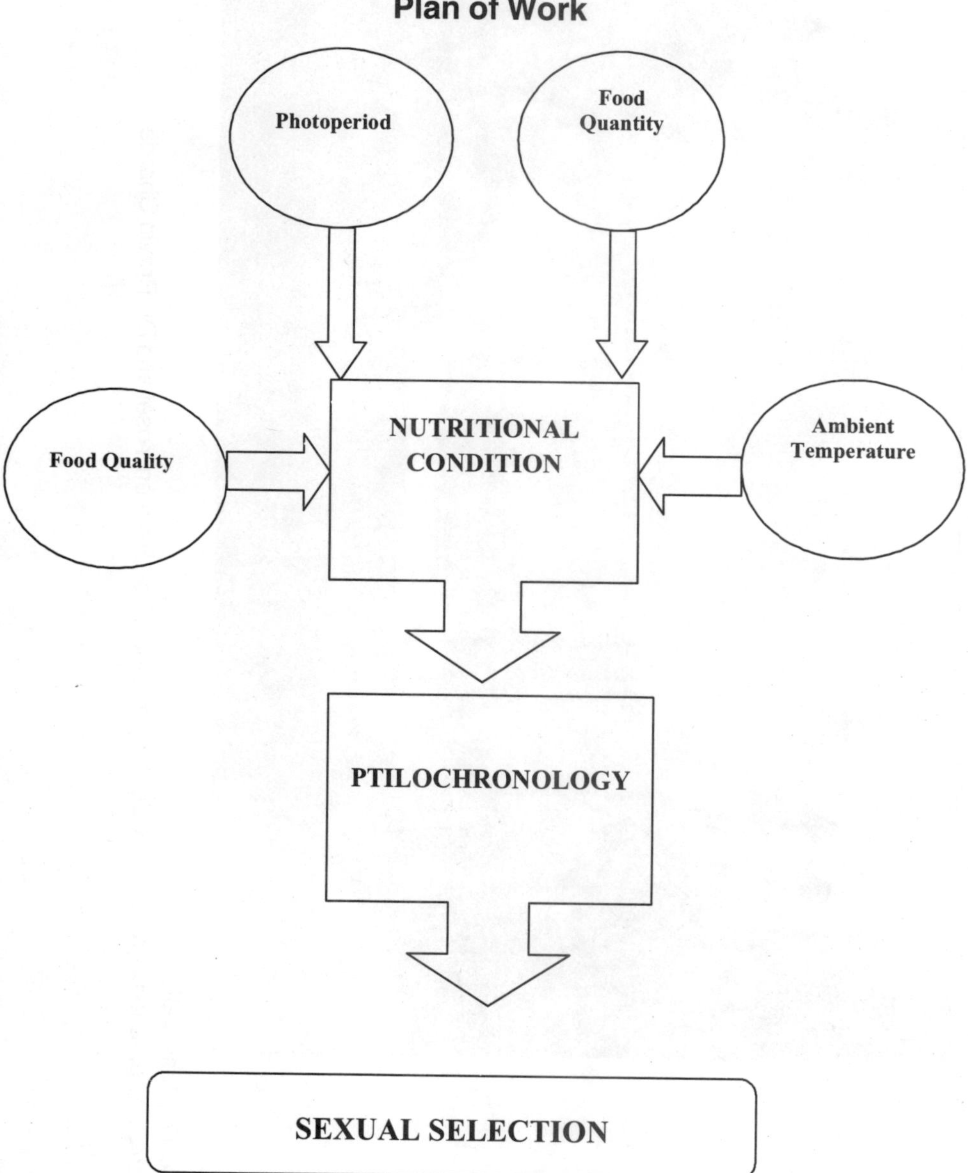

Figure 5.5: Schematic Representation of the Experimental Procedure.

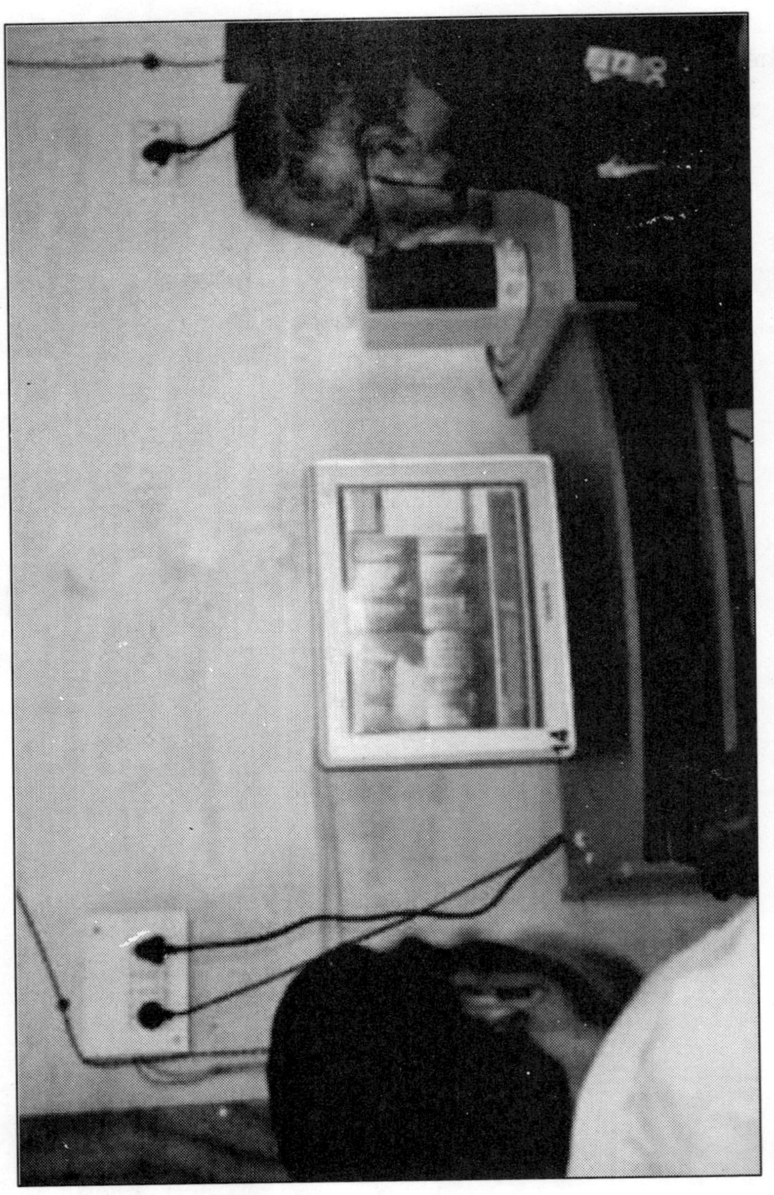

Plate 17: Observation.
Figure: Observation of recorded videos by menots Prof. Reuven Yosef and Dr. Pravin Charde.

Statistical Formulae

Statistical analysis was done by using the following formulae, referred from Fundamentals of Statistics (Srivastava and Srivastava 2003).

$$S^2 = \frac{\sum(d-\bar{d})^2}{n-1} \qquad (1)$$

where,

S: Standard deviation

d: Difference between variables x and y (x and y denotes the selection of control and experimental males respectively).

$\bar{d}$: Average of difference between variables x and y.

n: Number of pairs of control and experimental males.

$$t = \frac{\bar{d}}{s/\sqrt{n}} \qquad (2)$$

where,

t: Student t value

S: Standard deviation

$\bar{d}$: Average of difference between variables x and y (x and y denotes the selection of control and experimental males respectively).

n: Number of pairs of control and experimental males.

$$t = \frac{\bar{x}-\bar{y}}{\sqrt{s^2\left(\frac{1}{n_1}-\frac{1}{n_2}\right)}} \qquad (3)$$

where,

t: Student t value

S: Standard deviation

$\bar{x}$: Average daily growth bar width of control males

$\bar{y}$: Average daily growth bar width of experimental males.

n_1: Number of control males.

n_2: Number of experimental males.

$$S^2 = \frac{1}{n_1 + n_2 - 2}\left[\sum (x - \bar{x})^2 + \sum (y - \bar{y})^2\right] \quad (4)$$

where,

x: Daily growth bar width of control males.

y: Daily growth bar width of experimental males.

$\bar{x}$: Average daily growth bar width of control males

$\bar{y}$: Average daily growth bar width of experimental males.

n_1: Number of control males.

n_2: Number of experimental males.

Chapter 6
Results

Females when released for sexual selection trials were allowed to spend ca. 15 min (±5 min) in the central passage. During the first two minutes they walked the length of the passage looked at both males and then displayed their choice by either crouching near the chosen male or, if excited, hit the glass in order to try and reach the selected male.

Sexual selection decisions of females were not significantly affected by the quality of food (Student t test, t = 1.369, P = 0.102, n = 10) (Table 6A.2), nor by the extended light period of 17 hrs (t = 1.499, P = 0.104, n = 5). But photoperiod of nearly equal light and dark duration (L: D=13:11) significantly affected the selection decisions in females (t = 2.447, P = 0.035, n = 5) (Table 6B.2). The quantity of food including both 18 per cent (t = 0.514, P = 0.317, n = 5) and 16 per cent protein levels (t = 1.870, P = 0.067, n = 5) did not affect mate choice (Table 6C.2). In contrast, ambient temperature played a significant role in mate selection in female Japanese Quail (t = 2.572, P = 0.015, n = 10) (Table 6D.2).

The average growth bar widths of the central rectrices plucked of control (1.12±0.128) and manipulated males (1.08±0.128) suggested that ptilochronology is an independent function of food quality (Paired t-test, t = 0.714, P = 0.242, n = 10) (Table 6A.2).

Experiment 2 showed that photoperiod significantly affected the growth bar width, both when L: D=17:7 (control, 1.596±0.261 and experimental males, 1.226±0.261), t = 2.256, P = 0.270, n = 5 and when L: D=13:11 (control, 1.346±0.1208 and manipulated males 1.198±0.1208), t = 1.937, P = 0.044, n = 5 (Table 6B.2).

The alternate hypothesis has been accepted in the experiment 3, explaining that high protein quantity influenced the daily growth of the feathers (control, 1.172±0.1055 and manipulated males 0.966±0.1055 t = 3.0746, P = 0.007, n = 5, whereas low quantity of protein did not affect

Table 4.1: Initial Biometrics Including Body Mass (gms), Wing Length (mm), Tail Length (mm), Head Width (mm), Beak Width (mm), Beak Length (mm), Tarsus Length (mm) of Control and Experimental Japanese Quail After Completion of Experiments.

Sl.No.	*Males*	*Body Mass (gms)*	*Wing Length (mm)*	*Tail Length (mm)*	*Head Width (mm)*	*Beak Width (mm)*	*Beak Length (mm)*	*Tarsus Length (mm)*
1.	**C1**	178.4	105	41	19.37	4.51	19	38
2.	**C2**	156.2	119	40	18.81	4.86	17	36
3.	**C3**	166.8	116	42	19.51	4.22	19	37
4.	**C4**	159.6	109	41	18.54	4.71	17	35
5.	**C5**	167.4	112	40	19.73	5.49	17	36
6.	**C6**	161.6	111	38	19.12	5.08	18	36
7.	**C7**	170.4	109	39	18.48	5.12	17	37
8.	**C8**	155.2	109	36	16.65	4.89	19	37
9.	**C9**	158.6	110	40	17.52	4.56	19	34
10.	**C10**	163.2	112	42	18.12	5.01	21	36
11.	**E1**	164.6	113	35	18.40	4.96	19	38
12.	**E2**	178.6	118	42	18.62	4.32	18	37
13.	**E3**	220.2	117	41	18.71	5.74	21	41
14.	**E4**	162.2	111	43	18.63	5.02	19	38
15.	**E5**	183.4	105	40	17.25	3.89	20	39
16.	**E6**	161.4	98	40	17.72	4.62	19	38
17.	**E7**	193.4	110	42	19.48	4.16	21	35
18.	**E8**	186.6	119	35	16.37	4.73	21	39
19.	**E9**	178.8	110	32	17.75	5.44	21	38
20.	**E10**	168.8	111	35	15.75	5.29	19	36

the feather growth (control 1.03±0.1288 and experimental males, 1.016±0.1288) t = 0.1719, P = 0.433, n = 5 (Table 6C.2).

Similarly, ambient temperature did not affect the ptilochronology (control, 1.02±0.141 and manipulated males 0.97±0.141) of the birds (t = 0.794, P = 0.218, n = 10) (Table 4D.2).

The birds were first acclimatized for 10-12 days before procurement into the individually isolated cages. Out of the total randomly selected 20 birds were then placed in separate compartment as per the experimental arrangement. These birds were weighed and measured initially and data

Table 4.2: Final Biometrics Including Body Mass (gms), Wing Length (mm), Tail Length (mm), Head Width (mm), Beak Width (mm), Beak Length (mm), Tarsus Length (mm) of Control and Experimental Japanese Quail After Completion of Experiments.

Sl.No.	*Males*	*Body Mass (gms)*	*Wing Length (mm)*	*Tail Length (mm)*	*Head Width (mm)*	*Beak Width (mm)*	*Beak Length (mm)*	*Tarsus Length (mm)*
1.	**C1**	171.4	106	—	21.87	5.79	20	33
2.	**C2**	196.2	116	40	20.69	4.36	16	37
3.	**C3**	171.6	115	38	21.14	5.31	20	37
4.	**C4**	165	109	37	19.92	5.00	17	38
5.	**C5**	191.2	118	36	21.62	5.85	20	36
6.	**C6**	181.6	105	35	18.14	6.08	20	34
7.	**C7**	171.2	110	40	17.92	3.47	19	36
8.	**C8**	187.2	111	—	18.46	5.89	19	36
9.	**C9**	174.2	108	41	19.77	6.13	20	39
10.	**C10**	186.8	115	40	18.80	5.09	17	36
11.	**E1**	181.2	114	37	18.49	6.69	18	35
12.	**E2**	224.8	111	37	21.42	6.11	19	35
13.	**E3**	267.2	94	39	20.38	7.17	20	37
14.	**E4**	115.8	108	42	20.85	5.63	20	33
15.	**E5**	221.2	110	45	18.68	5.99	20	36
16.	**E6**	168.8	92	—	18.24	5.98	22	34
17.	**E7**	234.2	95	—	19.34	4.64	20	37
18.	**E8**	225.2	120	40	20.09	5.85	20	38
19.	**E9**	212.8	114	—	20.05	5.20	18	35
20.	**E10**	186.6	116	35	18.44	5.00	20	39

was recorded in terms of wing length, tail length, head width, beak width, beak length and tarsus length (Table 4.1).

These measurements were again recorded after completion of all the four sets of experiment (Table 4.2). Except one or two birds all were noted with an increased body mass. The length of wing remained nearly the same and the tail length was observed to decrease in majority of the birds. Some of the birds had also lost their tails. This may be due to the aberrant sexual behavior observed in these isolated males.

A. Experiment 1–FOOD QUALITY

As per the results of selection experiments, 100 per cent selection of experimental males fed with high quality diet was done in 5 pairs, 67 per cent selection of experimental males in 1 pair, 34 per cent selection in 3 pairs and 0 per cent selection in 1 pair. The females preferred the experimental males over the control males.

The feather growth bars of control males was found to be wider than that of experimental males (Plates 18, 19). The difference in between the growth bar widths of control and experimental males showed that the experimental males, fed with high quality food, developed comparitively narrower growth bars.

Table 4A.1: Body Mass (gms) of Control and Experimental Males during First Experiment.

Sl.No.	*Before Experimentation (04/10/09)*				*After Completion of Experimentation (12/10/09)*			
	Control Males	*Body Mass (gms)*	*Exptal Males*	*Body Mass (gms)*	*Control Males*	*Body Mass (gms)*	*Exptal Males*	*Body Mass (gms)*
1.	C1	178.4	E1	164.6	C1	181.4	E1	165.4
2.	C2	156.2	E2	178.6	C2	156.2	E2	179.2
3.	C3	166.8	E3	220.2	C3	172.6	E3	236.8
4.	C4	159.6	E4	162.2	C4	166.6	E4	166.4
5.	C5	167.4	E5	183.4	C5	172.2	E5	181.6
6.	C6	161.6	E6	161.4	C6	156.2	E6	159.6
7.	C7	170.4	E7	193.4	C7	166.6	E7	204.6
8.	C8	155.2	E8	186.6	C8	155.4	E8	198.2
9.	C9	158.6	E9	178.8	C9	157.2	E9	183.8
10.	C10	163..2	E10	168.8	C10	160.6	E10	172.8

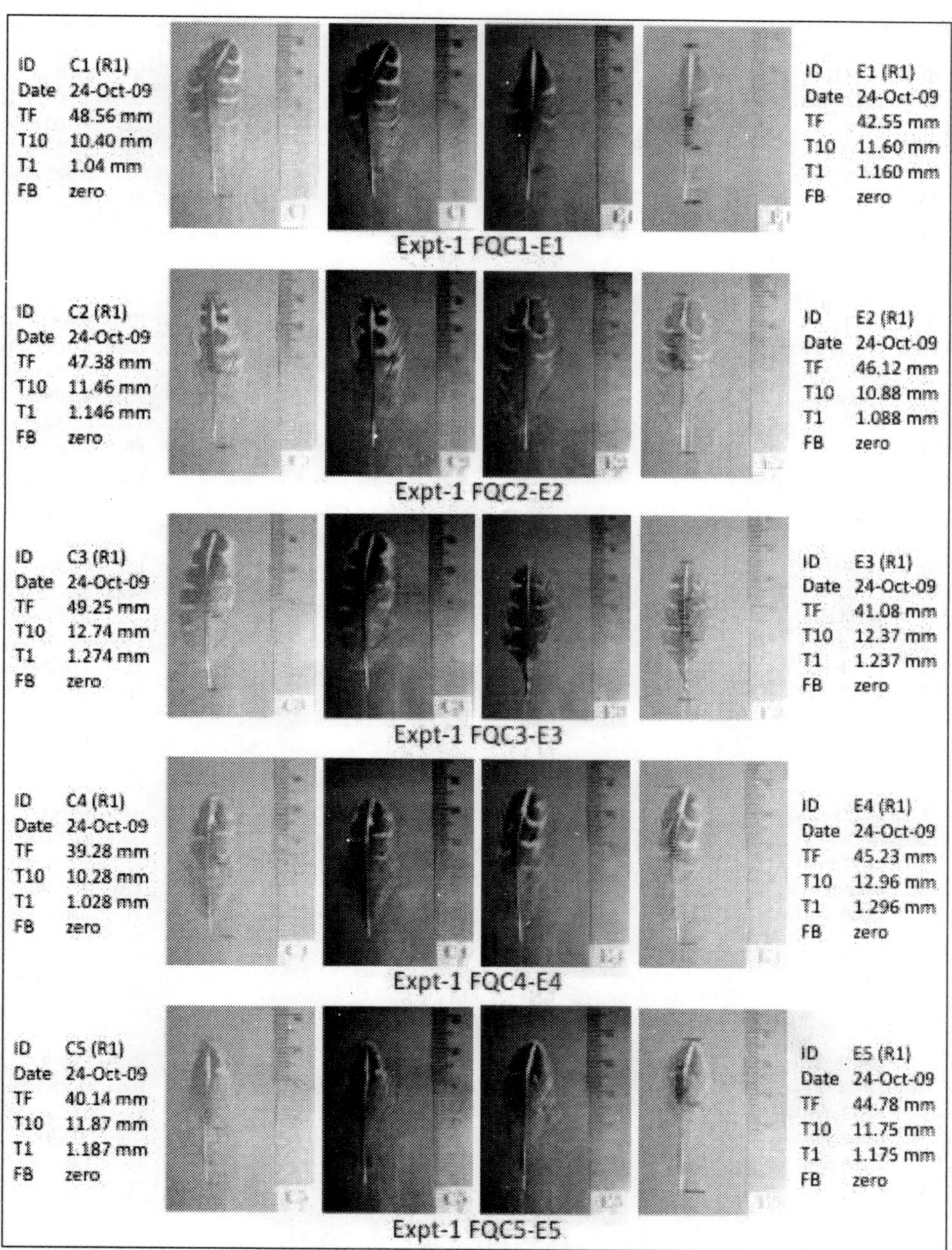

Plate 18: Expt 1–Food Quality (FQ).
Figure: Comparative ptilochronology of Control (C) and Experimental (E) Japanese Quail for Expt 1–FQ .

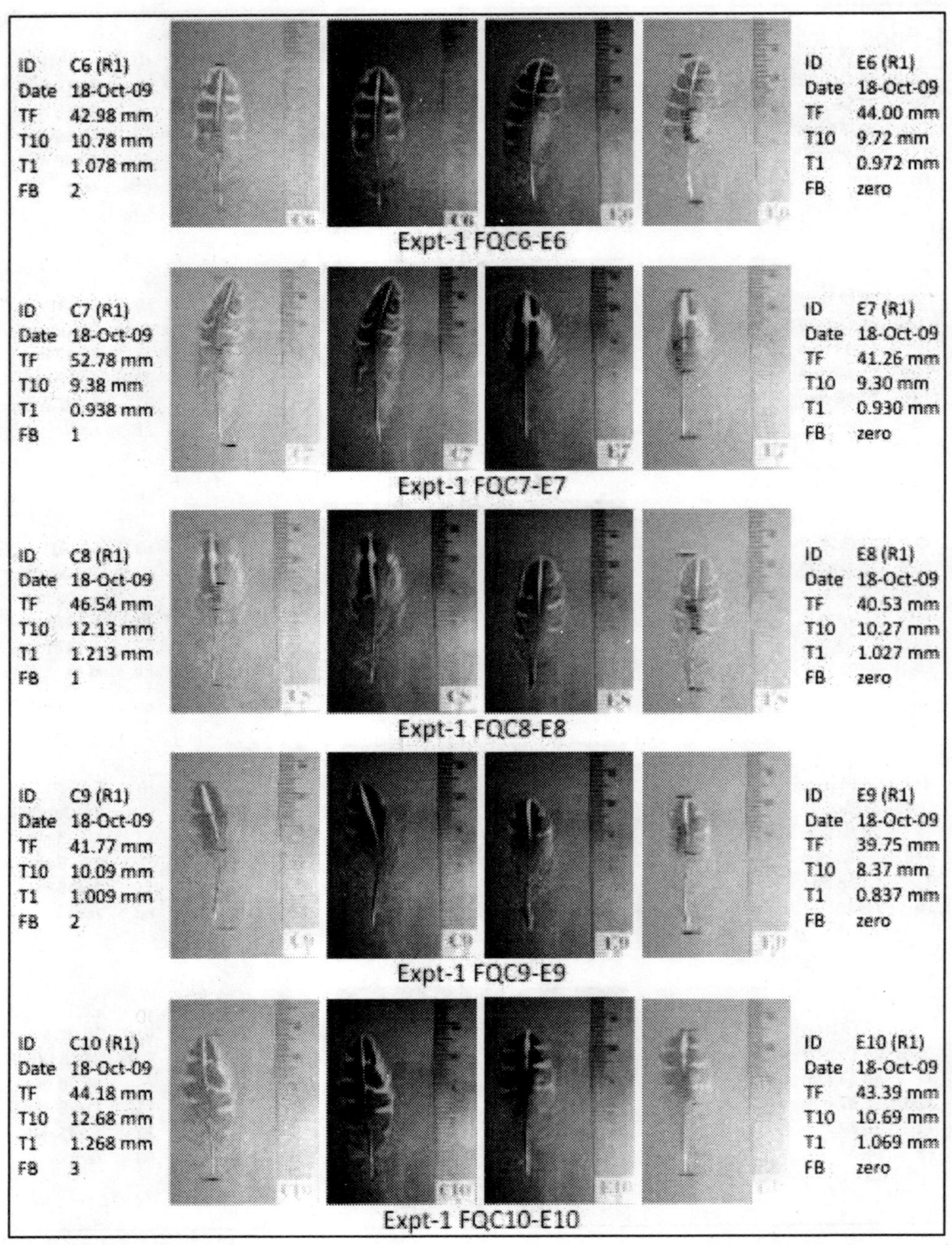

Plate 19: Expt 1–Food Quality (FQ).

Figure: Comparative ptilochronology of Control (C) and Experimental (E) Japanese Quail for Expt 1–FQ .

Table 4A.2: Results of selection experiment and average growth bar widths (X1) of right central rectrices (R1) of control (C) and experimental (E) males for experiment 1- Food Quality (Expt-1FQ).

Sl.No.	*Pairs of Males for Selection*	*Number of Females Introduced per Pair of Males (n=6)*		*Ptilochronology*				
		Control	*Experi-mental*	*ID*	*X1*	*ID*	*X1*	*Difference*
1.	C1-E1	0	6	**E1**	1.160	**C1**	1.04	0.12(+)
2.	C2-E2	0	6	**E2**	1.088	**C2**	1.146	0.058(-)
3.	C3-E3	4	2	**E3**	1.237	**C3**	1.274	0.037(-)
4.	C4-E4	4	2	**E4**	1.296	**C4**	1.028	0.268(+)
5.	C5-E5	6	0	**E5**	1.175	**C5**	1.187	0.012(-)
6.	C6-E6	0	6	**E6**	0.972	**C6**	1.078	0.106(-)
7.	C7-E7	0	6	**E7**	0.930	**C7**	0.938	0.008(-)
8.	C8-E8	2	4	**E8**	1.027	**C8**	1.213	0.186(-)
9.	C9-E9	4	2	**E9**	0.837	**C9**	1.009	0.172(-)
10.	C10-E10	0	6	**E10**	1.069	**C10**	1.268	0.199(-)

A. Expt 1–FQ

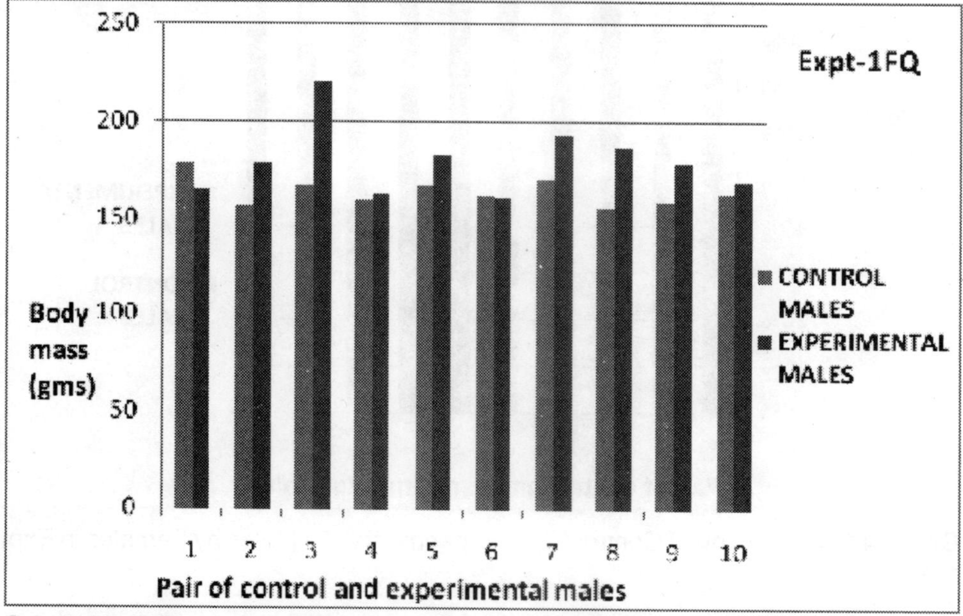

Graph 4A.1: Comparative Body Mass (gms) of Control (C) or Experimental (E) Males Prior to Expt 1–FQ.

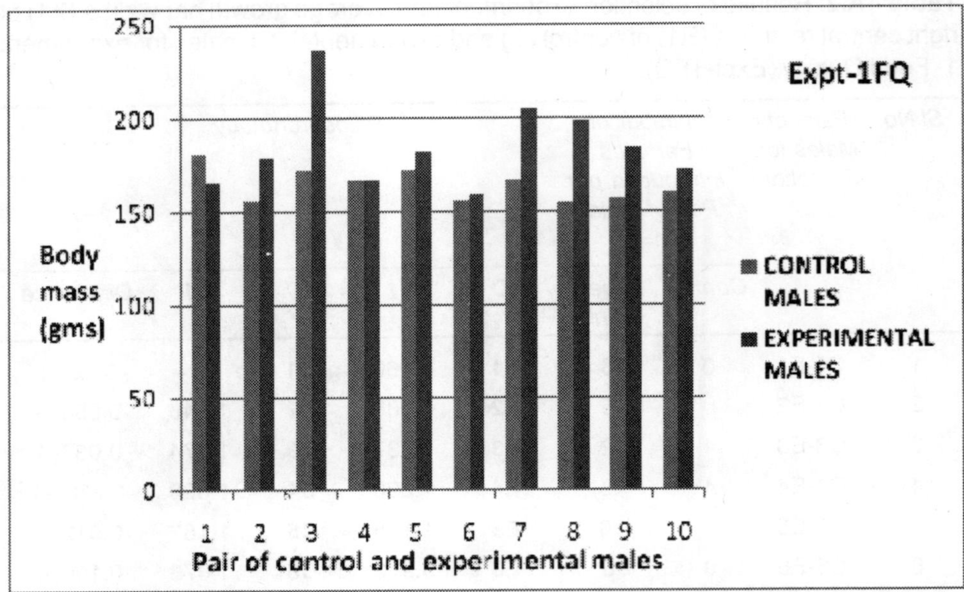

Graph 4A.2: Comparative Body Mass (gms) of Control (C) or Experimental (E) Males After Completion of Expt 1–FQ.

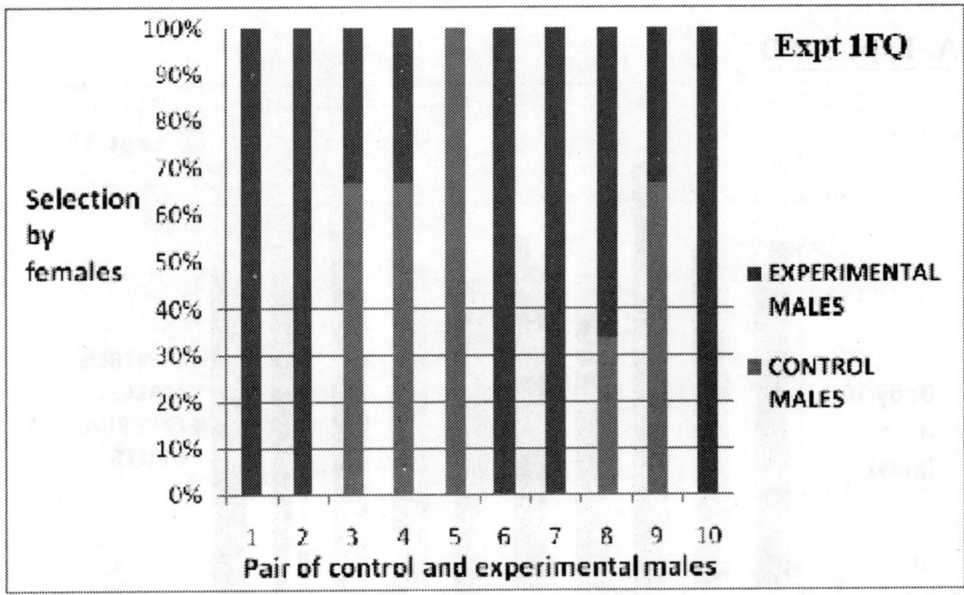

Graph 4A.3: Selection of Control (C) or Experimental (E) Males by Females in Expt 1–FQ.

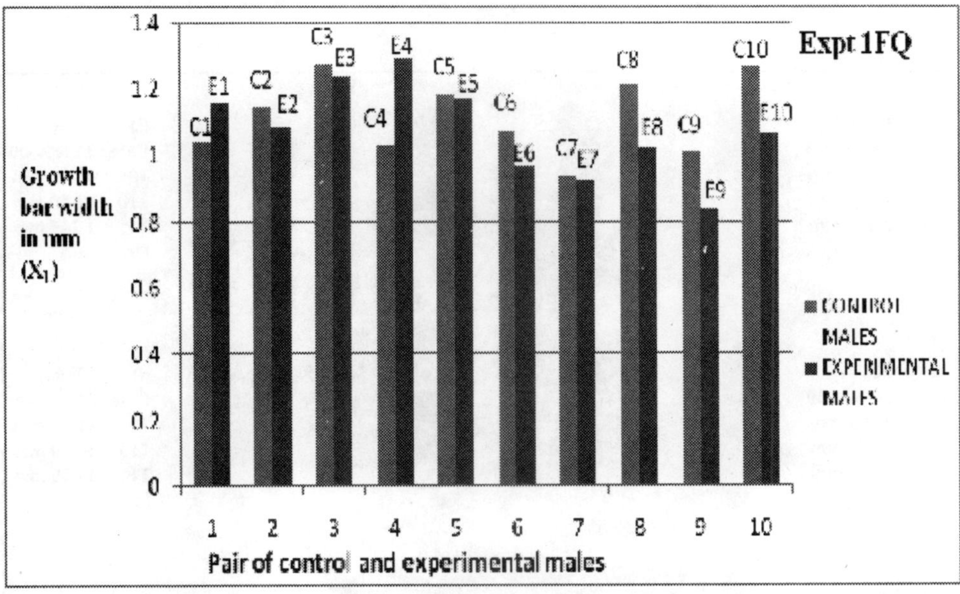

Graph 4A.4: Comparative Growth Bar Width of Control (C) and Experimental (E) Males in Expt 1–FQ.

B. Experiment 2–PHOTOPERIOD

The birds were weighed and their body mass were noted before starting and after completion of experiment.

Selection experiments were performed after completion of second experiment of photoperiod. In Expt 2a in which birds were exposed to longer light period, 67 per cent selction of control males were done by females in 4 pairs, and 34 per cent in 1 pair out of total five pairs. In Expt 2b where males were provided with shorter light duration, 100 per cent selection of control males were done in 1 pair, 67 per cent selection in 3 pairs and 50 per cent in 1 pair. The resuls indicate that females prefered control males over the experimental males in both shorter and longer light periods.

Ptilochronology of this experiment revealed that control males grew wider growth bars as compared to the experimental males in both the cases of shorter and longer light period.

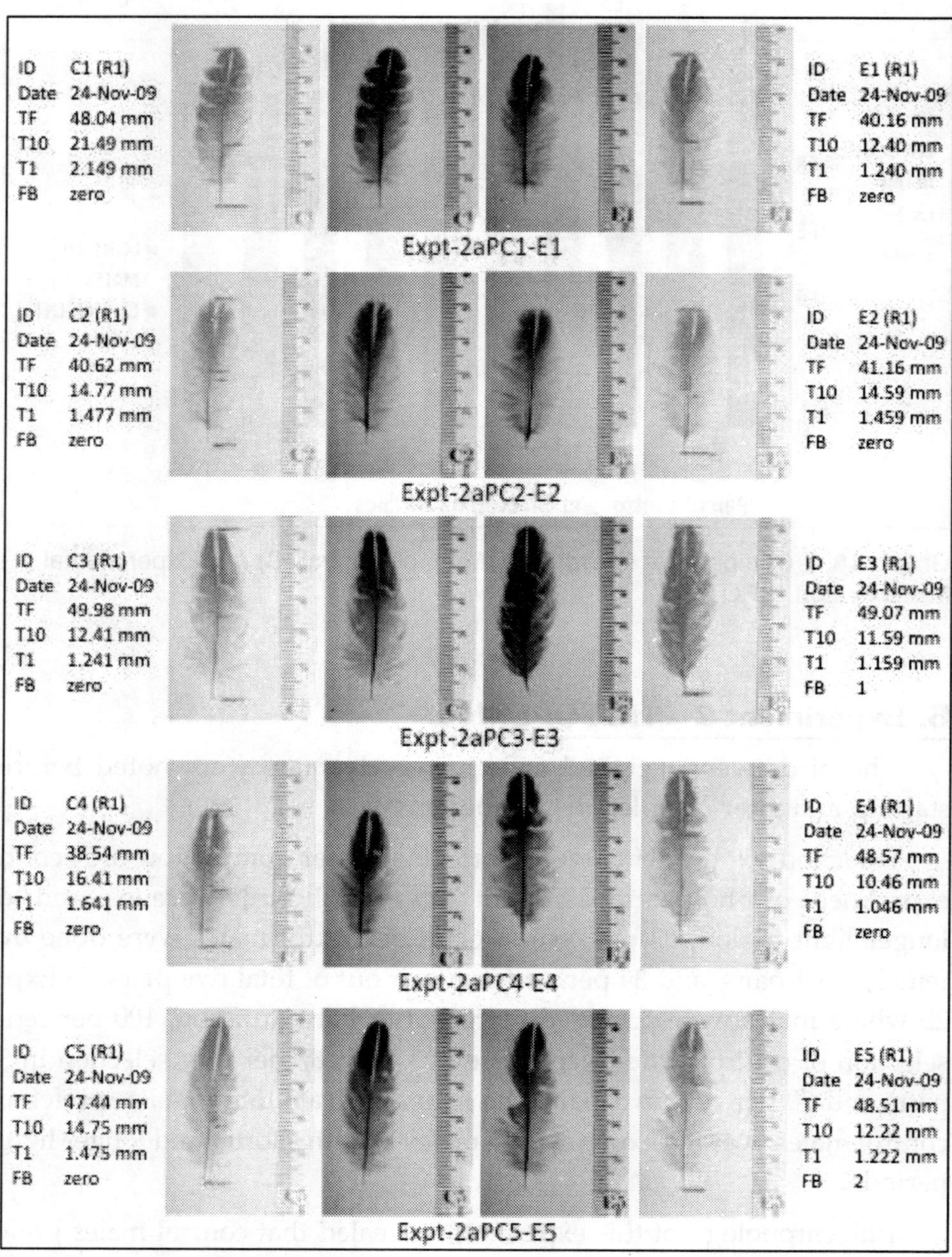

Plate 20: Expt 2a–Photoperiod (P).
Figure: Comparative ptilochronology of Control (C) and Experimental (E) Japanese Quail for Expt 2a–P.

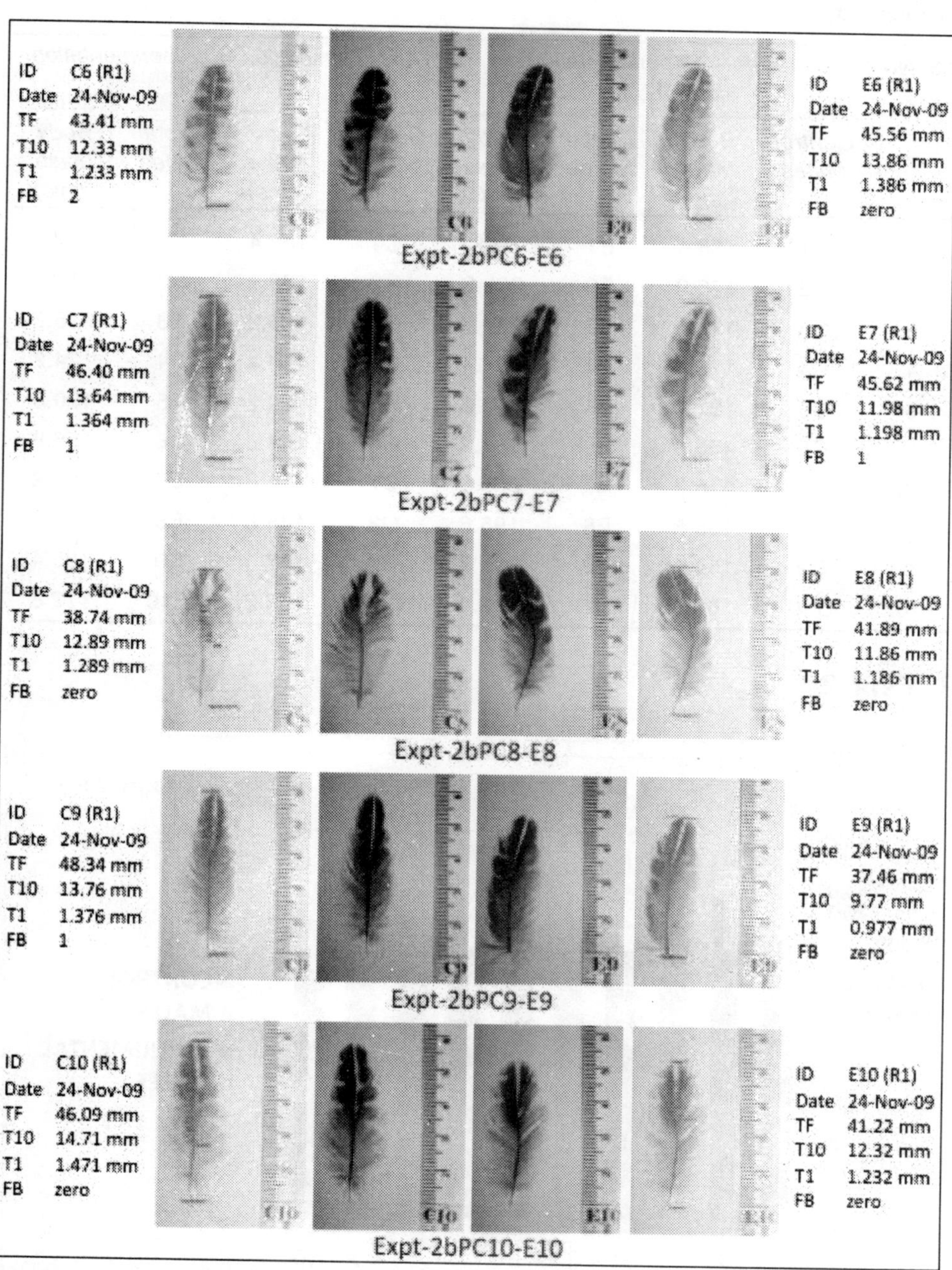

Plate 21: Expt 2b–Photoperiod (P).
Figure: Comparative ptilochronology of Control (C) and Experimental (E) Japanese Quail for Expt 2b–P.

Table 4B.1: Body Mass (gms.) of Control and Experimental Males during Second Experiment.

Sl.No.	*Before Experimentation (14/11/09)*				*After Completion of Experimentation (30/11/09)*			
	Control Males	*Body Mass (gms)*	*Exptal Males*	*Body Mass (gms)*	*Control Males*	*Body Mass (gms)*	*Exptal Males*	*Body Mass (gms)*
1.	C1	186.4	E1	168	C1	163.4	E1	167.2
2.	C2	175.2	E2	209.2	C2	183.2	E2	208
3.	C3	178.6	E3	250.8	C3	163.2	E3	225.2
4.	C4	173.4	E4	185.4	C4	158.4	E4	174.2
5.	C5	181.4	E5	182.6	C5	180.2	E5	190.6
6.	C6	156.4	E6	161.2	C6	164.6	E6	156.8
7.	C7	175.2	E7	208.2	C7	171.8	E7	208.2
8.	C8	155.6	E8	196.6	C8	148.8	E8	209.4
9.	C9	154.4	E9	189.8	C9	157.4	E9	185.4
10.	C10	160.8	E10	172.4	C10	163.2	E10	172.8

B. Expt 2P

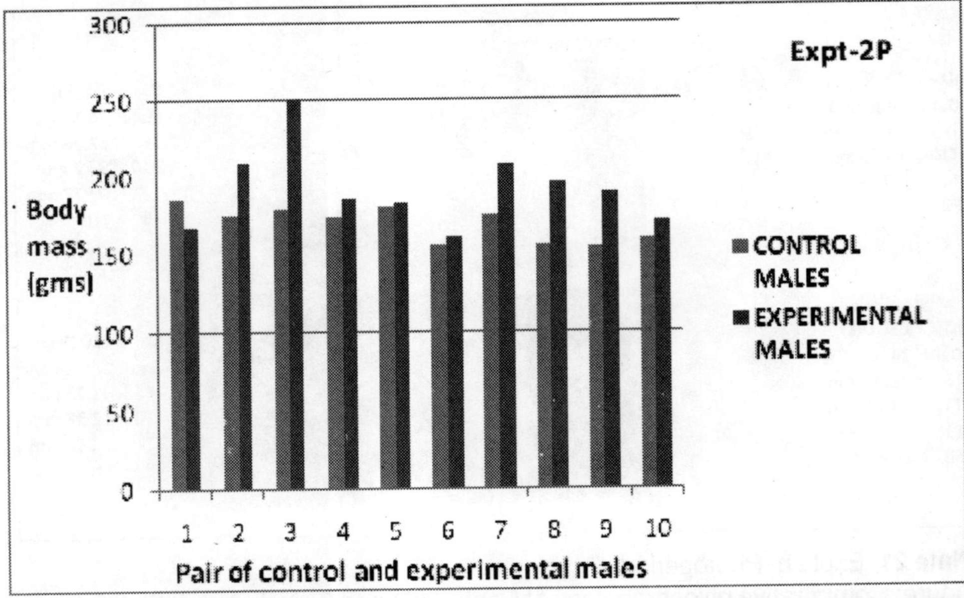

Graph 4B.1: Comparative Body Mass (gms) of Control (C) or Experimental (E) Males Prior to Expt-2P.

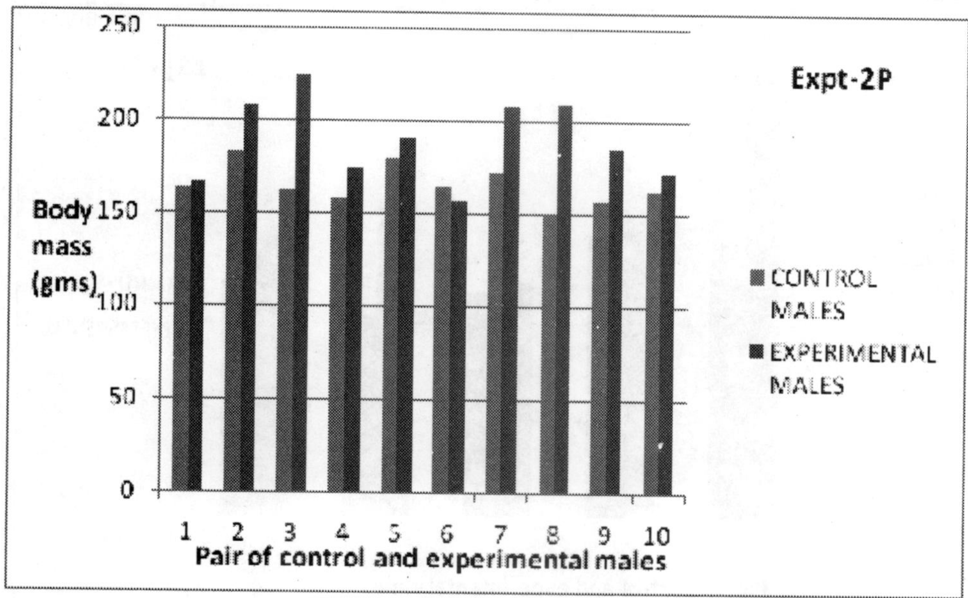

Graph 4B.2: Comparative Body Mass (gms) of Control (C) or Experimental (E) Males After Completion of Expt-2P.

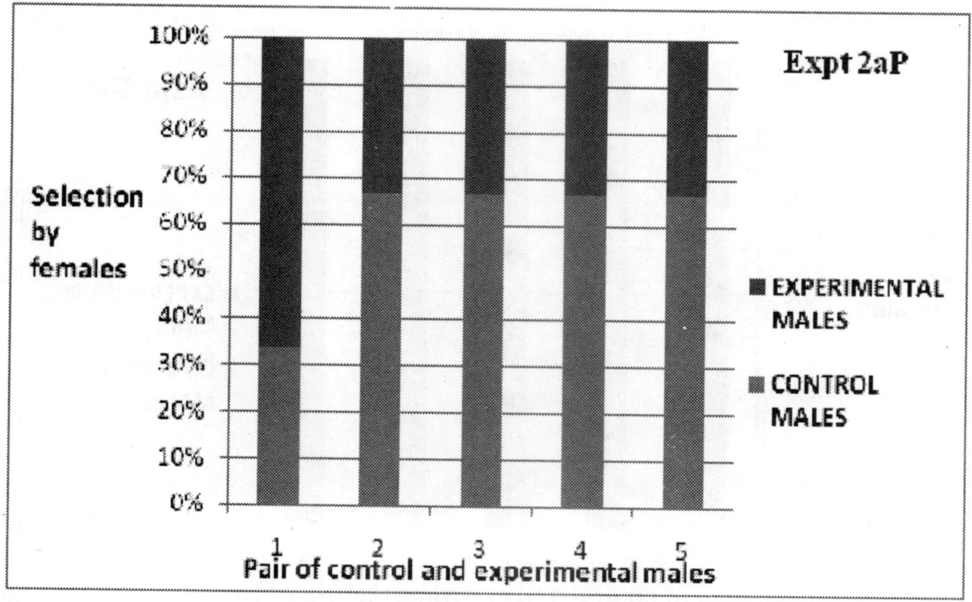

Graph 4B.3: Selection of Control (C) or Experimental (E) Males by Females in Expt-2aP.

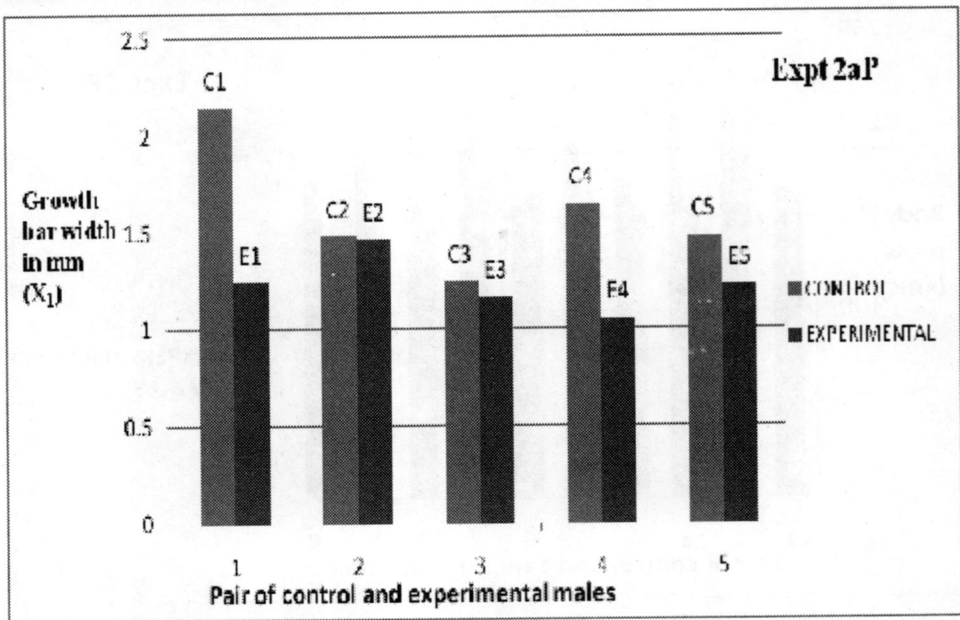

Graph 4B.4: Comparative Growth Bar Width of Control (C) and Experimental (E) Males in Expt-2aP.

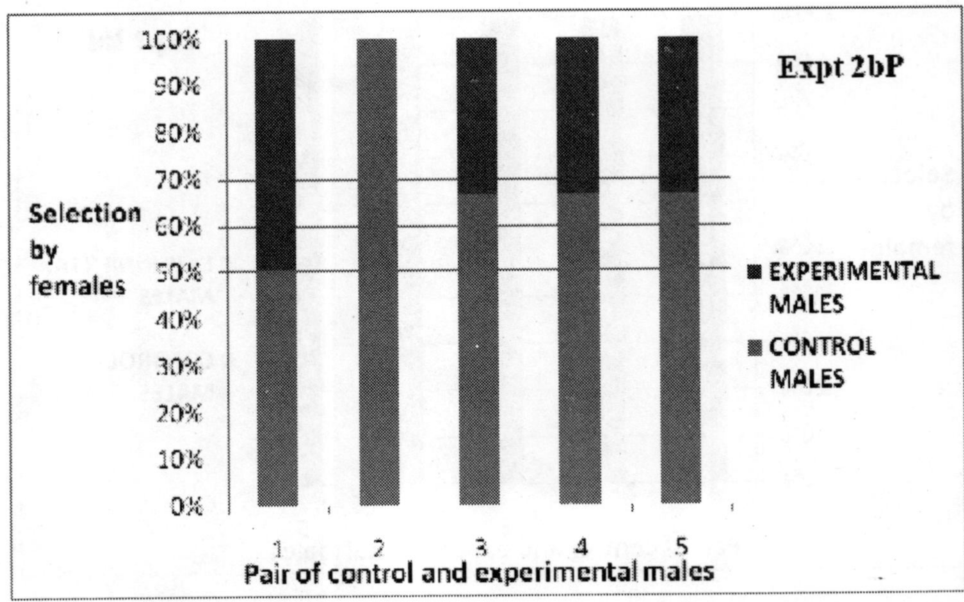

Graph 4B.5: Selection of Control (C) or Experimental (E) Males by Females in Expt-2bP.

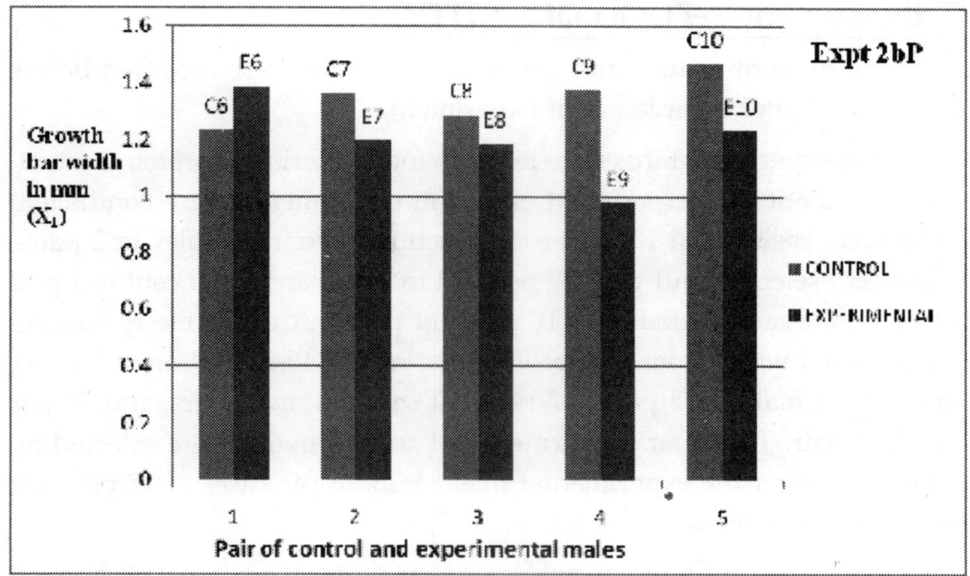

Graph 4B.6: Comparative Growth Bar Width of Control (C) and Experimental (E) Males in Expt-2bP.

Table 4B.2: Results of selection experiment and average growth bar widths (X1) of right central rectrices (R1) of control (C) and experimental (E) males for experiment 2a and 2b - Photoperiod (Expt-2aP and Expt 2bP).

Sl.No.	*Pairs of Males for Selection*	*Number of Females Introduced per Pair of Males (n=6)*		*Ptilochronology*				
		Control	*Experi-mental*	*ID*	*X1*	*ID*	*X1*	*Difference*
				2a = 17:7				
1.	C1-E1	2	4	**E1**	1.240	**C1**	2.149	0.909(-)
2.	C2-E2	4	2	**E2**	1.459	**C2**	1.477	0.018(-)
3.	C3-E3	4	2	**E3**	1.159	**C3**	1.241	0.082(-)
4.	C4-E4	4	2	**E4**	1.046	**C4**	1.641	0.595(-)
5.	C5-E5	4	2	**E5**	1.222	**C5**	1.475	0.253(-)
				2b = 13:11				
6.	C6-E6	3	3	**E6**	1.386	**C6**	1.233	0.153(+)
7.	C7-E7	6	0	**E7**	1.198	**C7**	1.364	0.166(-)
8.	C8-E8	4	2	**E8**	1.186	**C8**	1.289	0.103(-)
9.	C9-E9	4	2	**E9**	0.977	**C9**	1.376	0.399(-)
10.	C10-E10	4	2	**E10**	1.232	**C10**	1.471	0.239(-)

C. Experiment 3–FOOD QUANTITY

Body mass of control and experimental males were recorded before starting and after completion of experiment.

In this experiment birds were fed with food differing in protein content. After completion of experiment, selection experiments were conducted. These were resulted in 100 per cent selection of control males in 2 pairs, 67 per cent selection in 1 pair, 37 per cent in 1 pair and 0 per cent in 1 pair out of total five pairs fed with 18 per cent protein diet. In the remaining five pairs fed with 16 per cent protein diet resulted in selection of 100 per cent control males in 2 pairs, 67 per cent selection in 2 pairs, and 34 per cent in 1 pair. This clearly indicated that control males were selected by the females over the experimental males in both the cases of 18 per cent and 16 per cent protein diet.

Growth bar width of control males was wider than the experimental males in all the pairs fed with 18 per cent and 16 per cent protein diet (Plates 22, 23).

Table 4C.1: Body Mass (gms) of Control and Experimental Males during Third Experiment.

Sl.No.	*Before Experimentation (10/01/10)*				*After Completion of Experimentation (01/02/10)*			
	Control Males	*Body Mass (gms)*	*Exptal Males*	*Body Mass (gms)*	*Control Males*	*Body Mass (gms)*	*Exptal Males*	*Body Mass (gms)*
1.	C1	160	E1	185.2	C1	169	E1	190.6
2.	C2	186.4	E2	228.2	C2	183.2	E2	220.6
3.	C3	166.6	E3	223.2	C3	172.4	E3	250.8
4.	C4	156.4	E4	173.2	C4	158.2	E4	172.2
5.	C5	187.2	E5	227.6	C5	194.4	E5	218.6
6.	C6	171.2	E6	165.2	C6	172.6	E6	168.6
7.	C7	178	E7	216.2	C7	175.8	E7	218.6
8.	C8	175.8	E8	222	C8	185	E8	216.8
9.	C9	162.8	E9	196	C9	166.8	E9	201.4
10.	C10	176.6	E10	176	C10	175	E10	180.2

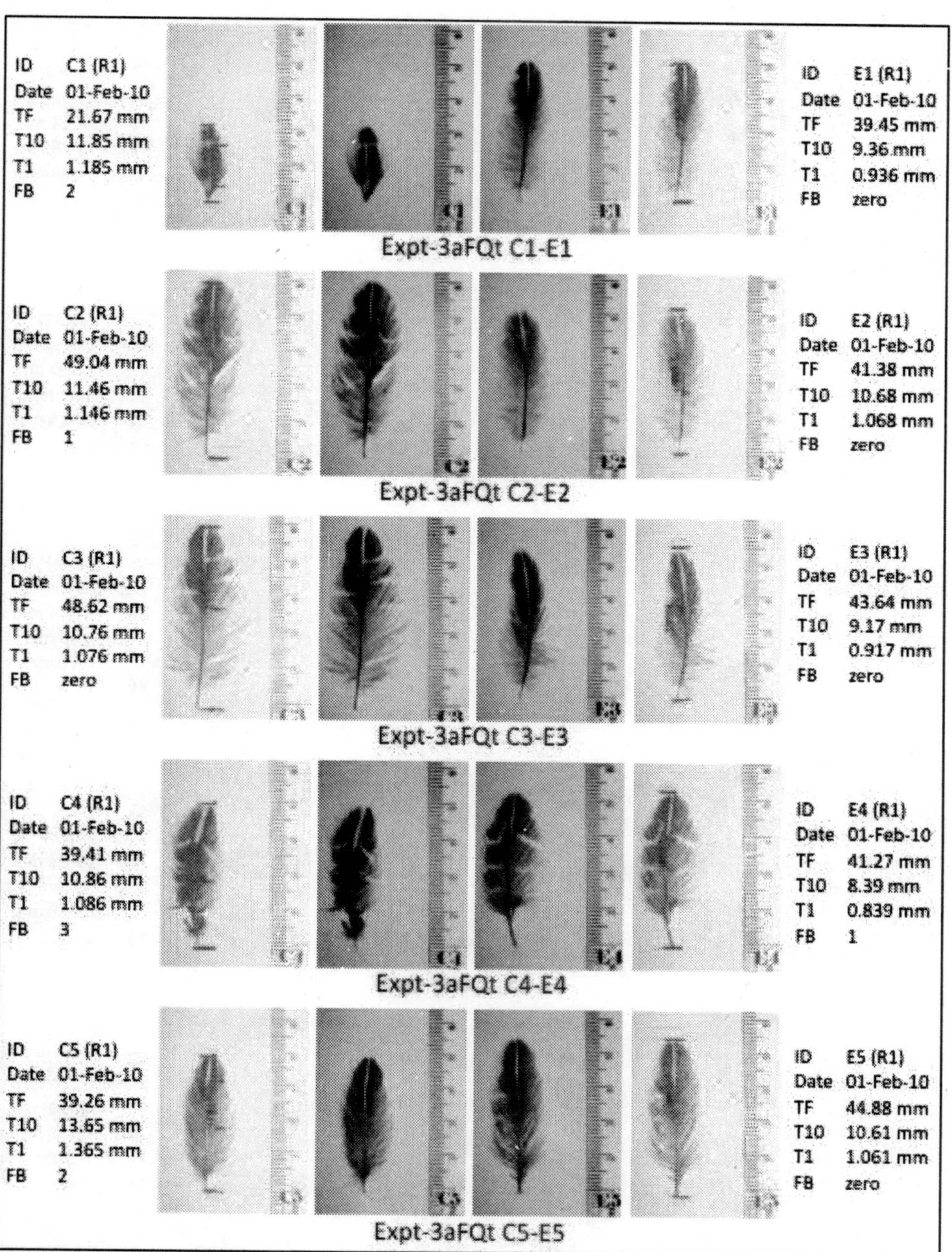

Plate 22: Expt 3a–Food Quantity (FQt).

Figure: Comparative ptilochronology of Control (C) and Experimental (E) Japanese Quail for Expt 3a-FQt.

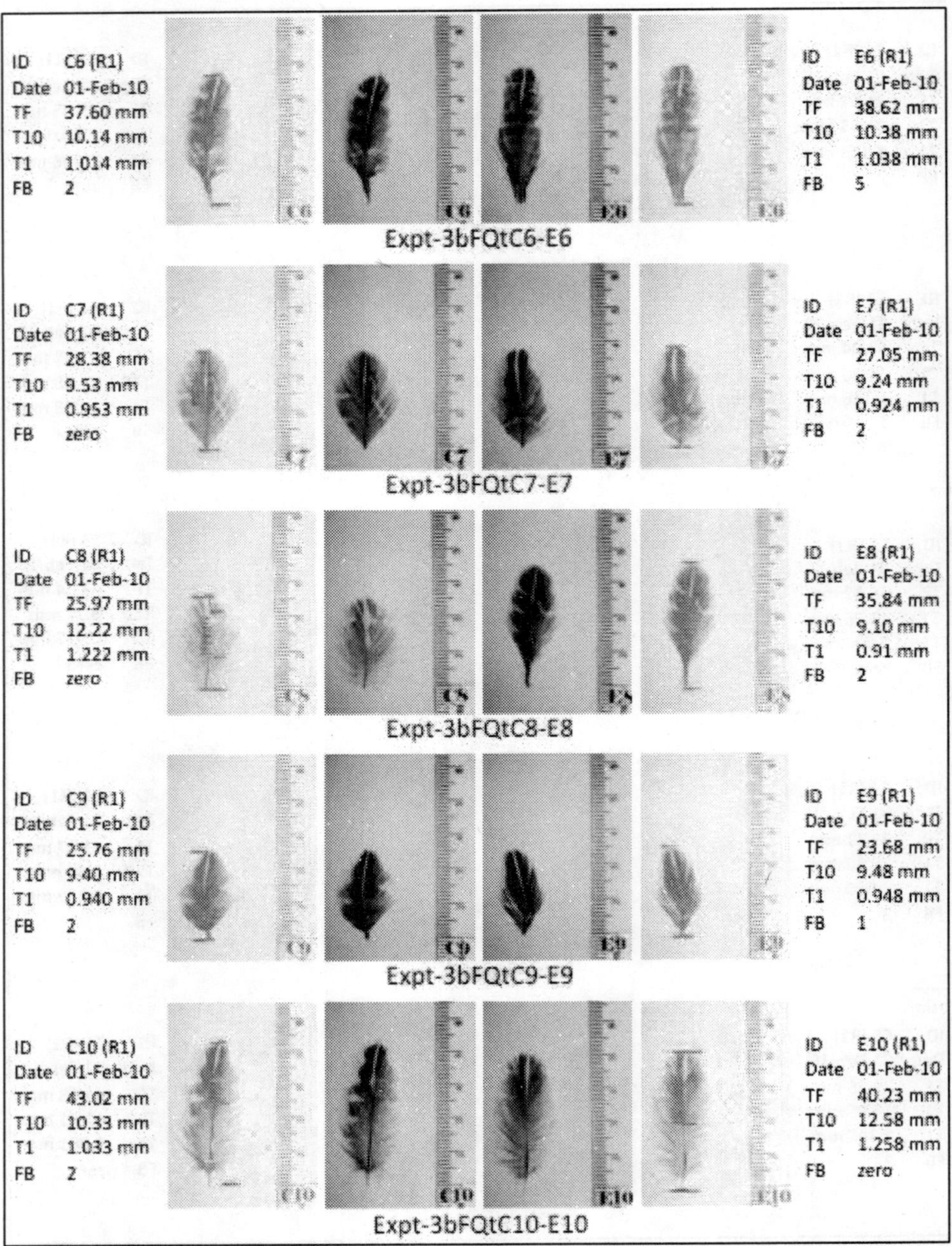

Plate 23: Expt 3b–Food Quantity (FQt).
Figure: Comparative ptilochronology of Control (C) and Experimental (E) Japanese Quail for Expt 3b–FQt.

C. Expt 3–FQt

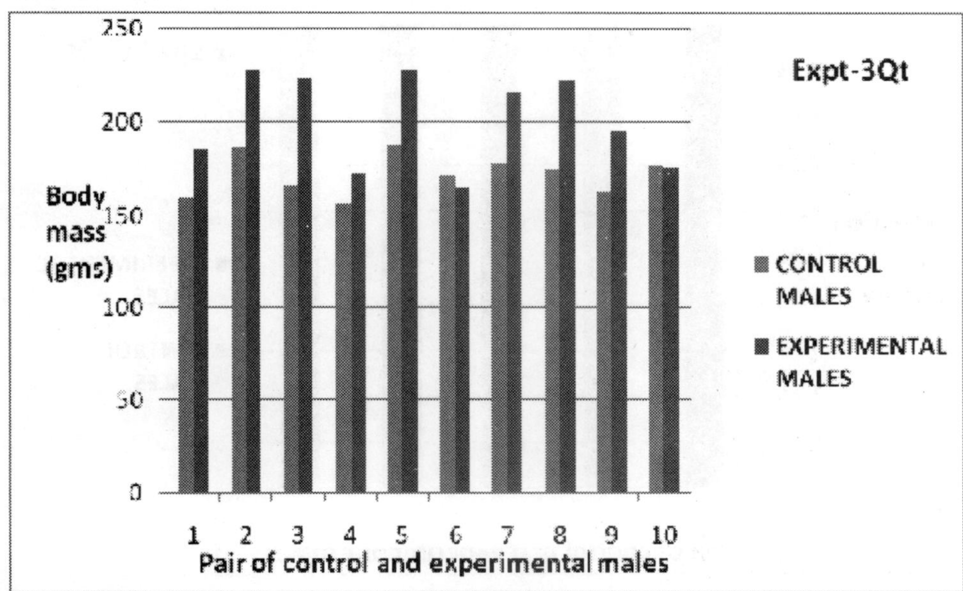

Graph 4C.1: Comparative Body Mass (gms) of Control (C) or Experimental (E) Males Prior to Expt 3–FQt.

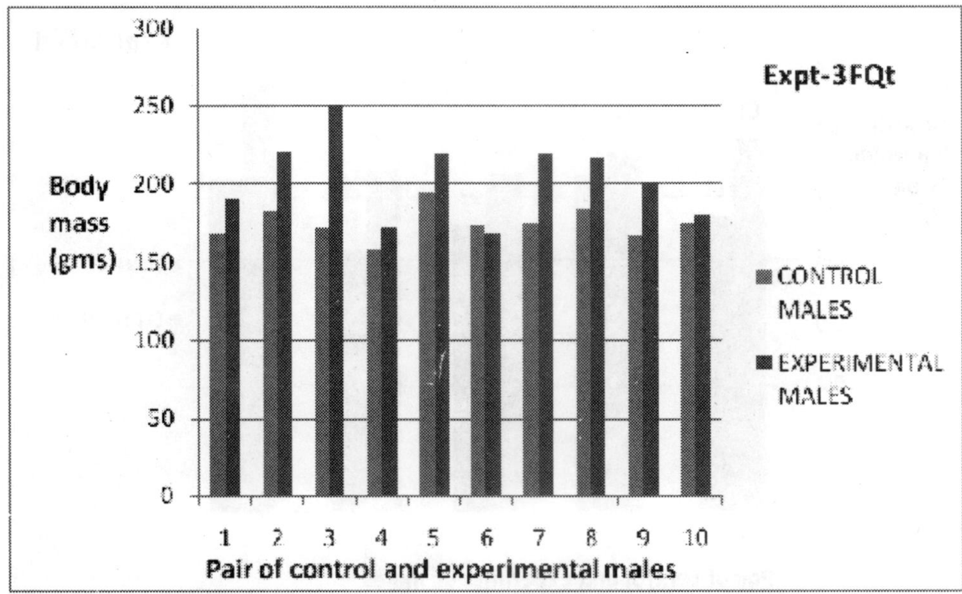

Graph 4C.2: Comparative Body Mass (gms) of Control (C) or Experimental (E) Males After Completion of Expt 3–FQt.

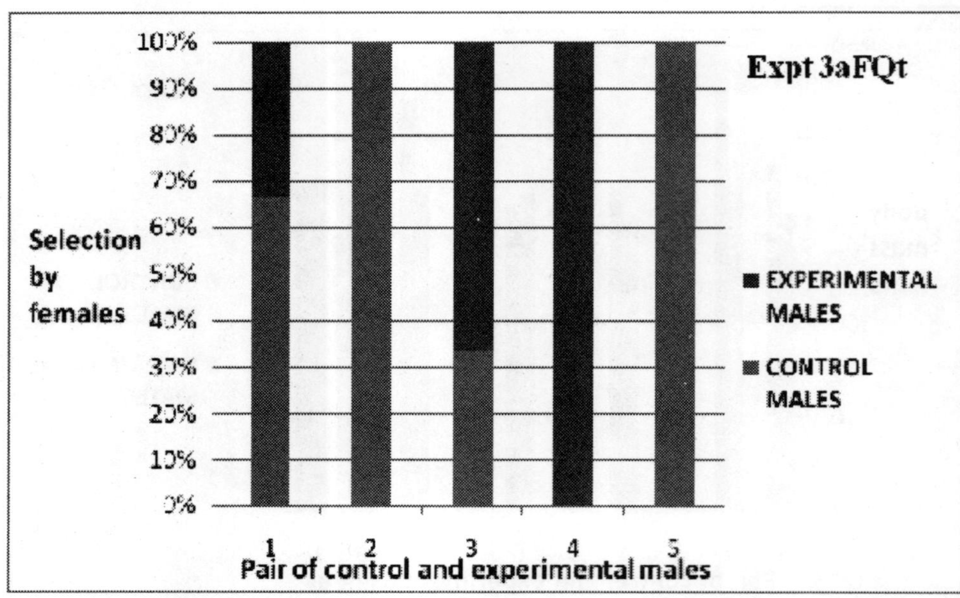

Graph 4C.3: Selection of Control (C) or Experimental (E) Males by Females in Expt 3a–FQt.

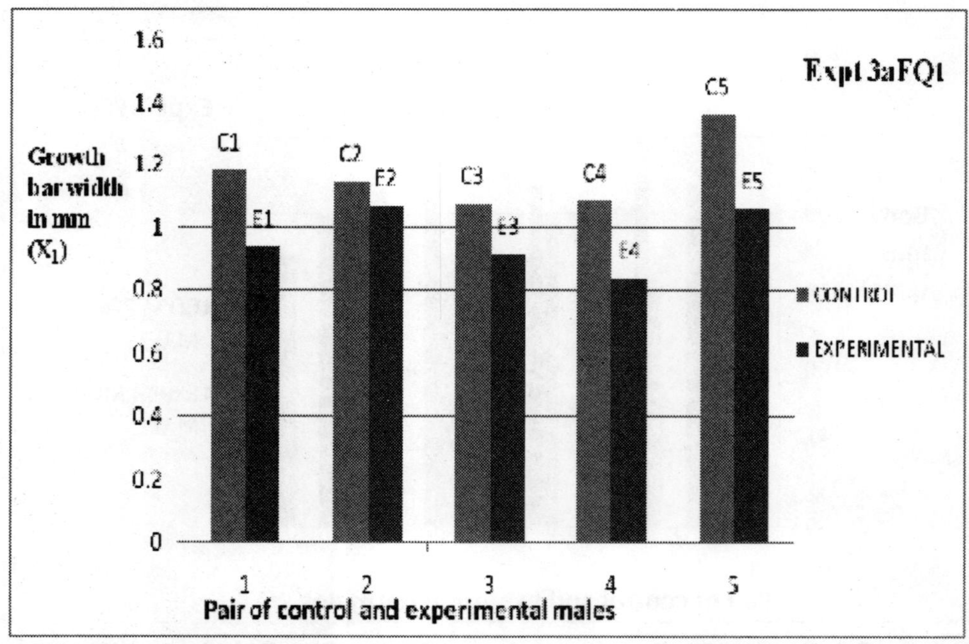

Graph 4C.4: Comparative Growth Bar Width of Control (C) and Experimental (E) Males in Expt 3a–FQt.

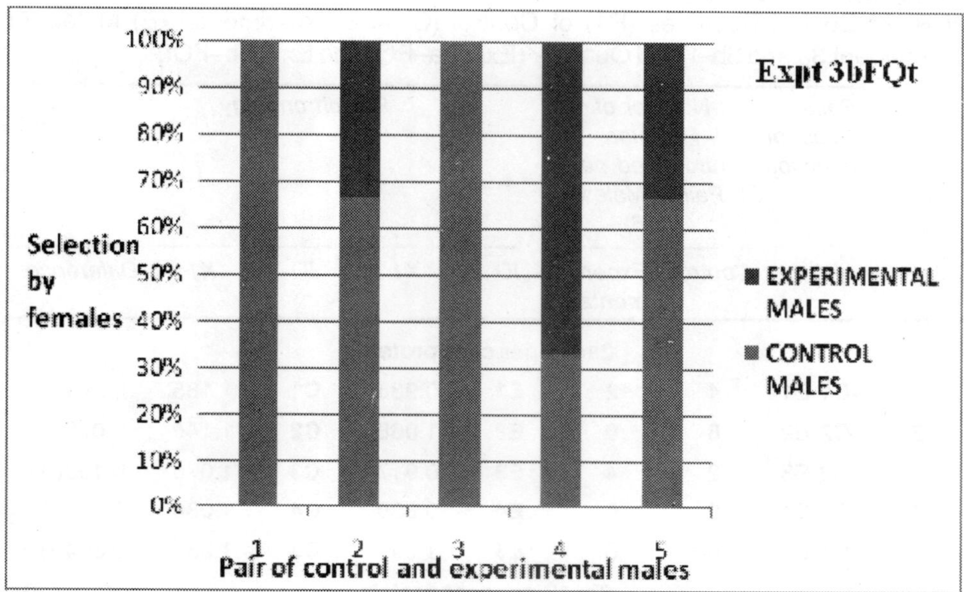

Graph 4C.5: Selection of control (C) or experimental (E) males by females in Expt 3b–FQt.

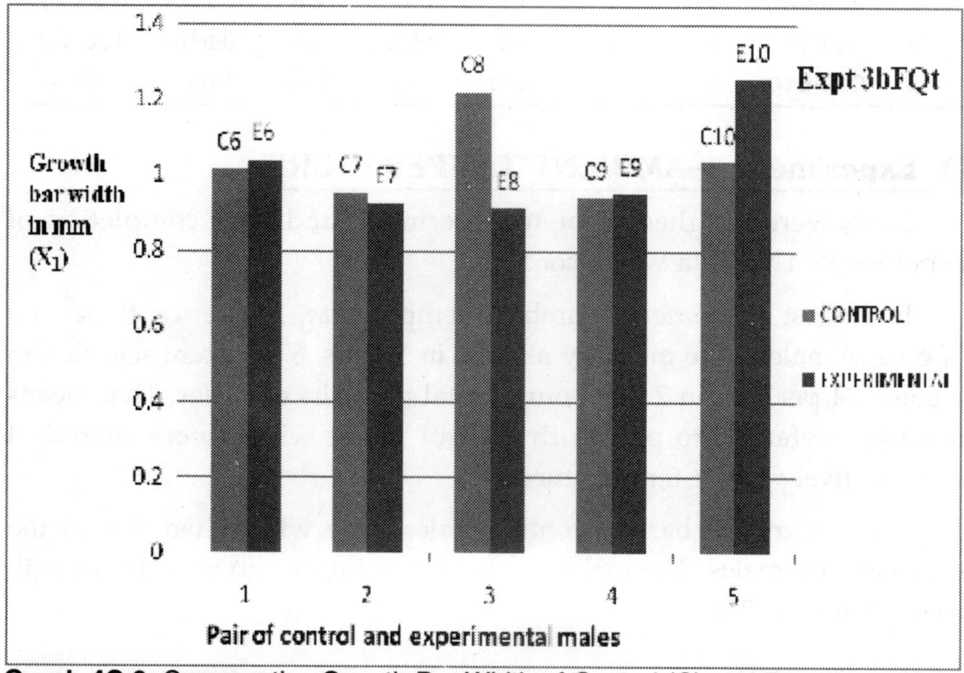

Graph 4C.6: Comparative Growth Bar Width of Control (C) and Experimental (E) Males in Expt 3b–FQt.

Table 4C.2: Results of Selection Experiment and Average Growth Bar Widths (X1) of Right Central Rectrices (R1) of Control (C) and Experimental (E) Males for Experiment 3a and 3b–Food Quantity (Expt 3a–FQt and Expt 3b–FQt).

Sl.No.	*Pairs of Males for Selection*	*Number of Females Introduced per Pair of Males (n=6)*		*Ptilochronology*				
		Control	*Experi-mental*	*ID*	*X1*	*ID*	*X1*	*Difference*
				3a=18 per cent protein				
1	C1-E1	4	2	**E1**	0.936	**C1**	1.185	0.249(-)
2	C2-E2	6	0	**E2**	1.068	**C2**	1.146	0.078(-)
3	C3-E3	2	4	**E3**	0.917	**C3**	1.076	0.159(-)
4	C4-E4	0	6	**E4**	0.839	**C4**	1.086	0.247(-)
5	C5-E5	6	0	**E5**	1.061	**C5**	1.365	0.304(-)
				3b=16 per cent protein				
6	C6-E6	6	0	**E6**	1.038	**C6**	1.014	0.024(+)
7	C7-E7	4	2	**E7**	0.924	**C7**	0.953	0.029(-)
8	C8-E8	6	0	**E8**	0.91	**C8**	1.222	0.312(-)
9	C9-E9	2	4	**E9**	0.948	**C9**	0.940	0.008(+)
10	C10-E10	4	2	**E10**	1.258	**C10**	1.033	0.225(+)

D. Experiment 4–AMBIENT TEMPERATURE

Birds were weighed prior to experiment and after completion of experiment. This data was recorded.

In the last experiment of ambient temperature, 100 per cent selection of control males were made by females in 3 pairs, 67 per cent selection in 5 pairs, 34 per cent in 2 pairs out of total ten pairs of males. That means females preferred to pair with control males which were reared at comparatively lower temperature.

Feather growth bars of control males were wider than that of the experimental males. The males in cool temperatures grew wider growth bars (Plates 24, 25).

Table 4D.1: Body Mass (gms) of Control and Experimental Males during Fourth Experiment.

Sl.No.	*Before Experimentation (08/02/10)*				*After Completion of Experimentation (07/03/10)*			
	Control Males	*Body Mass (gms)*	*Exptal Males*	*Body Mass (gms)*	*Control Males*	*Body Mass (gms)*	*Exptal Males*	*Body Mass (gms)*
1.	C1	169.2	E1	189.6	C1	171.4	E1	181.2
2.	C2	182.6	E2	221	C2	196.2	E2	224.8
3.	C3	171.2	E3	257.8	C3	171.6	E3	267.2
4.	C4	155.6	E4	172.6	C4	165	E4	115.8
5.	C5	195.6	E5	218	C5	191.2	E5	221.2
6.	C6	176.6	E6	171.2	C6	181.6	E6	168.8
7.	C7	173.8	E7	217.4	C7	171.2	E7	234.2
8.	C8	187.2	E8	222	C8	187.2	E8	225.2
9.	C9	166.8	E9	204.2	C9	174.2	E9	212.8
10.	C10	187.6	E10	179.4	C10	186.8	E10	186.6

Table 4D.2: Results of Selection Experiment and Average Growth Bar Widths (X1) of Right Central Rectrices (R1) of Control (C) and Experimental (E) Males for Experiment 4–Ambient Temperature (Expt–4AT).

Sl.No.	*Pairs of Males for Selection*	*Number of Females Introduced per Pair of Males (n=6)*		*Ptilochronology*				
		Control	*Experi-mental*	*ID*	*X1*	*ID*	*X1*	*Difference*
1.	C1-E1	4	2	**E1**	0.883	**C1**	0.931	0.048(-)
2.	C2-E2	4	2	**E2**	0.895	**C2**	1.089	0.194(-)
3.	C3-E3	4	2	**E3**	0.816	**C3**	1.173	0.357(-)
4.	C4-E4	6	0	**E4**	0.846	**C4**	1.262	0.416(-)
5.	C5-E5	6	0	**E5**	1.095	**C5**	0.929	0.834(-)
6.	C6-E6	4	2	**E6**	1.162	**C6**	0.839	0.323(+)
7.	C7-E7	2	4	**E7**	0.808	**C7**	1.020	0.212(-)
8.	C8-E8	4	2	**E8**	0.877	**C8**	0.959	0.082(-)
9.	C9-E9	6	0	**E9**	1.088	**C9**	0.930	0.158(+)
10.	C10-E10	2	4	**E10**	1.230	**C10**	1.054	0.176(+)

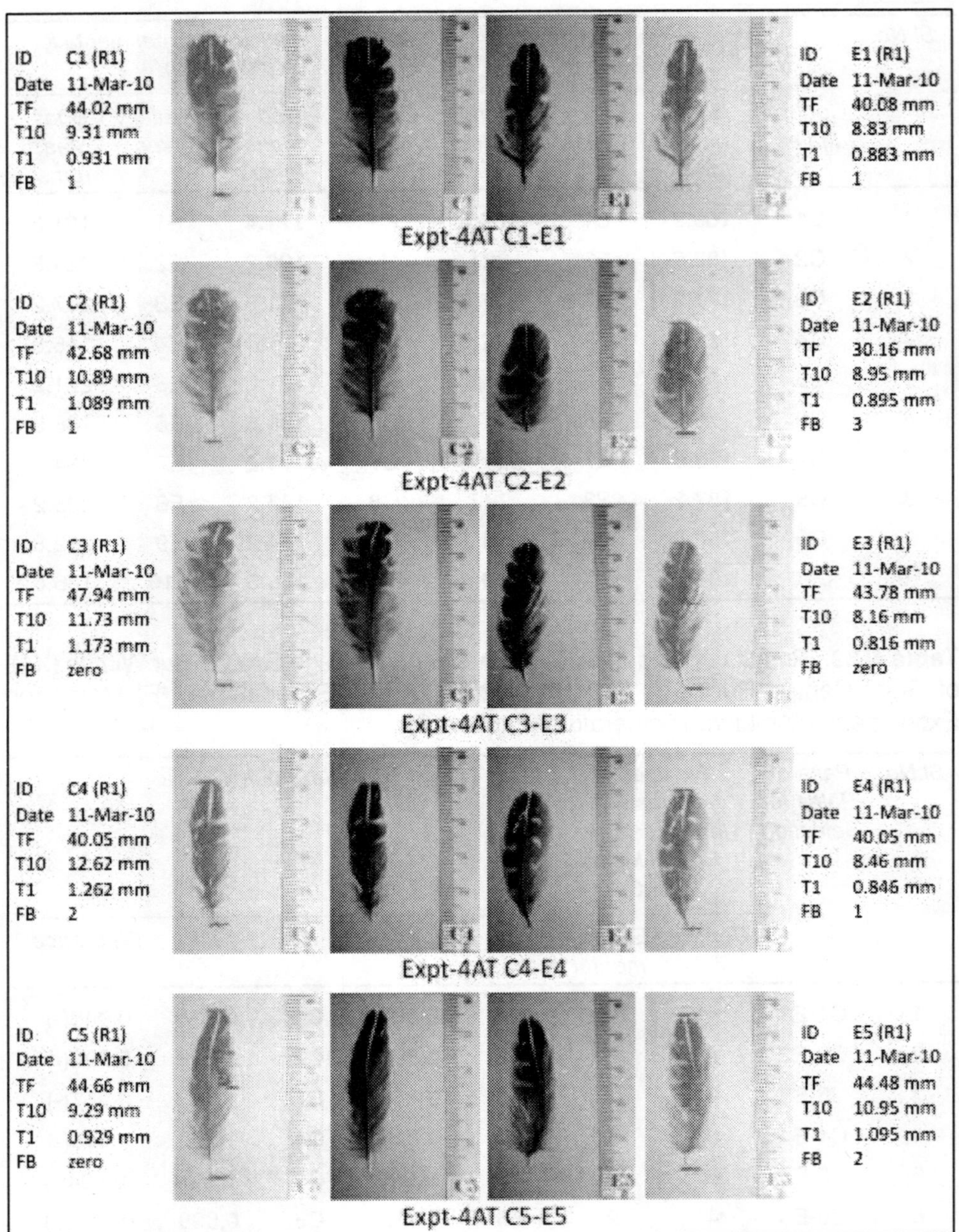

Plate 24: Ambient Temperature (AT).

Figure: Comparative ptilochronology of Control (C) and Experimental (E) Japanese Quail for Expt 4-AT.

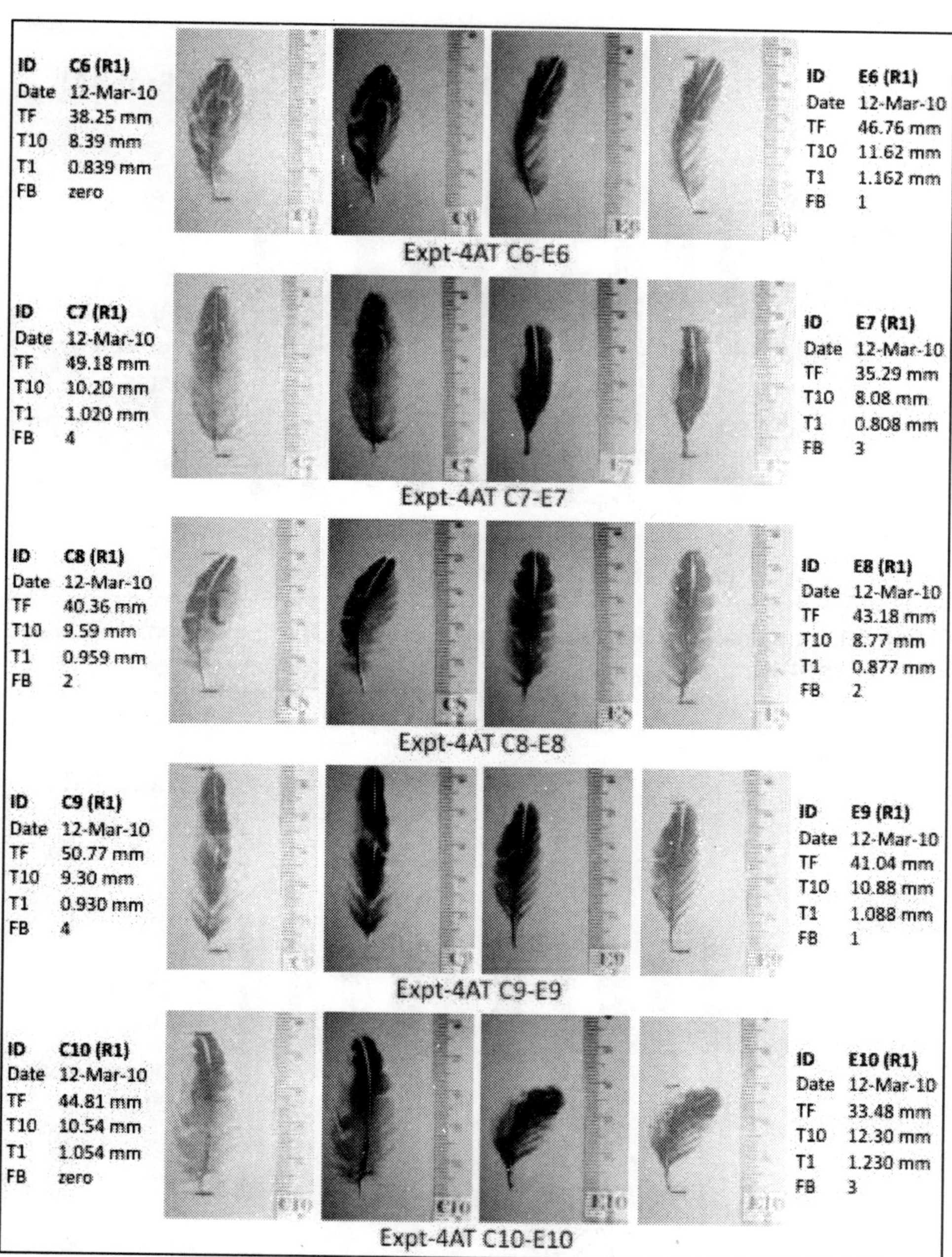

Plate 25: Ambient Temperature (AT).

Figure: Comparative ptilochronology of Control (C) and Experimental (E) Japanese Quail for Expt 4-AT.

D. Expt 4AT

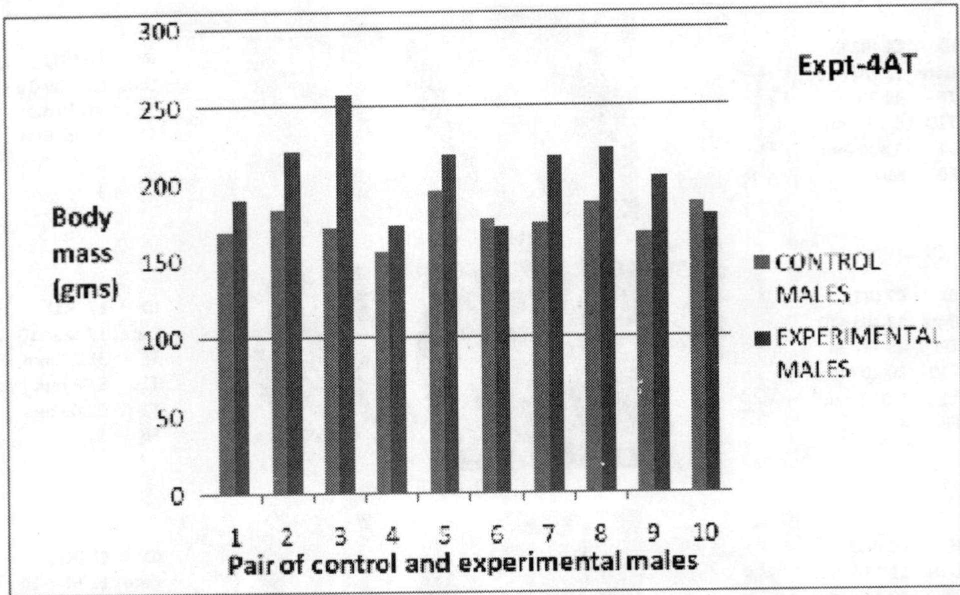

Graph 4D.1: Comparative Body Mass (gms) of Control (C) or Experimental (E) Males Prior to Expt 4–AT.

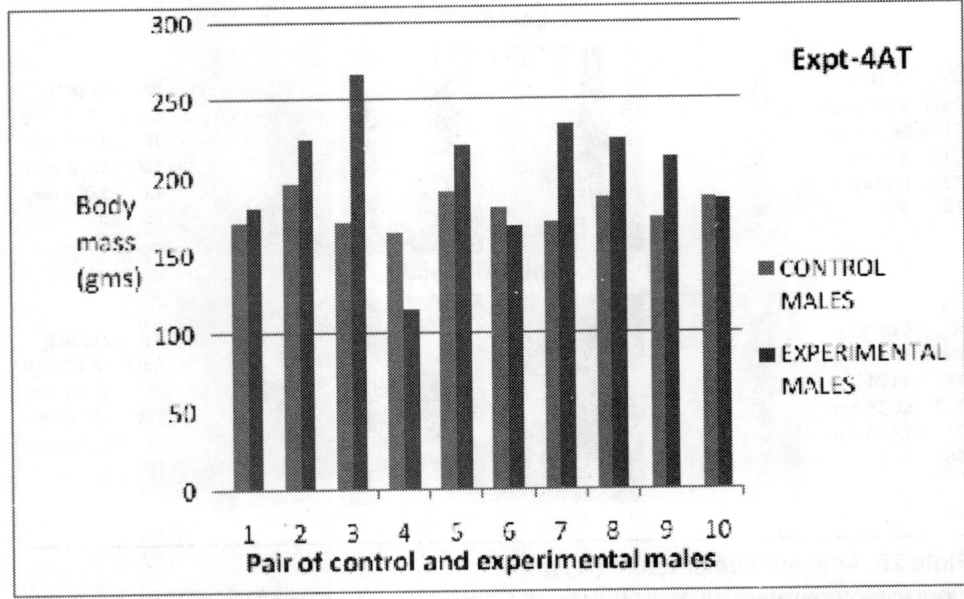

Graph 4D.2: Comparative Body Mass (gms) of Control (C) or Experimental (E) Males After Completion of Expt 4–AT.

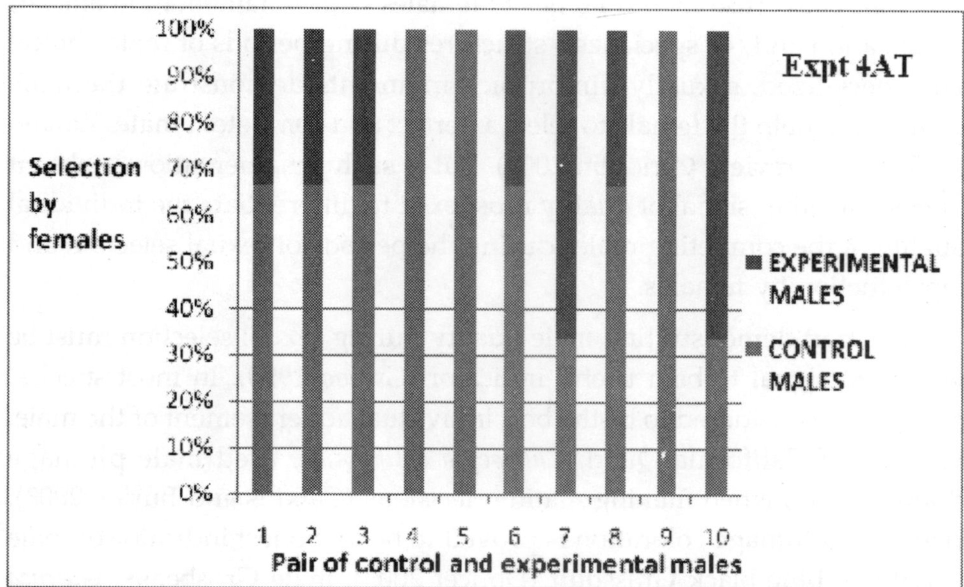

Graph 4D.3: Selection of Control (C) or Experimental (E) Males by Females in Expt 4–AT.

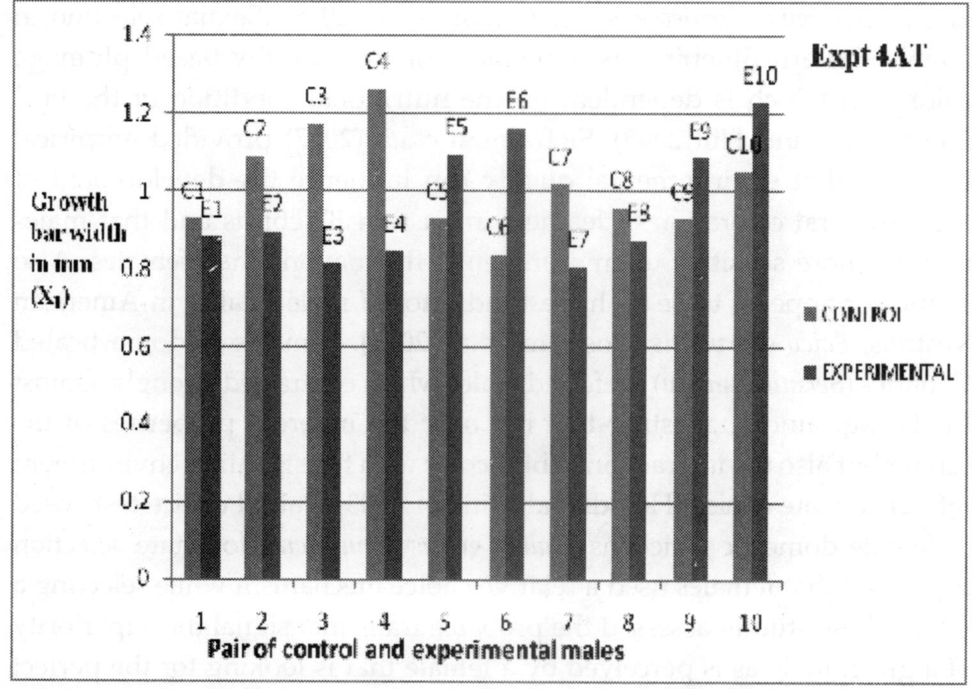

Graph 4D.4: Comparative Growth Bar Width of Control (C) and Experimental (E) Males in Expt 4–AT.

Males advertise their qualities to females with ornamentation and/or vocalization and/or specialized structures during periods of mate choice. The specialized, sexually dimorphic, ornaments demonstrate the male quality and help the female to select a perfect and competent mate (Zahavi 1975 but see review Candolin 2003). But if such ornamentation is absent then some other signal of quality must exist to differentiate the individual quality of the competing males during the periods of sexual selection and reproduction by females.

The trait demonstrating male quality during sexual selection must be its honest signal to be a useful indicator (Grafen 1990). In most species, plumage is considered to be the best individual advertisement of the male. Females of California Quail, *Callipepla californica,* used male plumage characteristics when making mating decisions (Calkins and Burley 2003). Structural plumage coloration is proved to be an honest indicator of male quality in Blue-black Grassquit (Doucet 2002); male Grosbeaks, *Guiraca caerulea* (Keyser *et al.,* 1999); in Eastern Bluebirds, *Sialia sialis* (Siefferman and Hill 2003); in female House Finches (Hill 1990) and in Florida Scrub-Jay, *Aphelocoma caerulescens* (Siefferman *et al.,* 2008). Sexual selection in female Eastern Bluebirds is dependent on structurally based plumage coloration which is dependent on the nutritional condition of the bird (Siefferman and Hill 2005). Siefferman *et al.* (2007) provided empirical evidence that environmental quality can influence the development of the structural coloration of feathers in Eastern Bluebirds and that males may be more sensitive to environmental fluctuations than females. Also skin color appears to be an honest indicator of male quality in American Kestrels, *Falco sparverius* (Bostrom *et al.,* 2006). Females in domesticated Canaries (*Serinus canaria*) preferred males which contrasted strongly against the background suggesting that not only the inherent properties of the sender but also the interaction of bird color with the signaling environment influence mate choice (Heindl and Winkler 2003). Paired choice tests used in female domestic Chickens, *Gallus gallus domesticus,* for mate selection suggested that females used a relative choice mechanism while selecting a mate. These studies assessed the probable traits that signal the superiority of a given male as is perceived by a female that is looking for the perfect mate for the forthcoming breeding effort.

Experiment 1–Food Quality

Hill and Montgomerie (1994) found that in House Finches there was a significant positive relation between the growth rate of a male's tail feathers and the brightness of his plumage, suggesting that males growing brighter feathers are in better nutritional condition. Hence, because the Japanese Quail that we studied are monomorphic, we considered the relative nutritional condition of each male to be the trait defining mate choice by females for our experiment. We hypothesized that better nutritional condition resulted from the intake of high quality food which resulted in the growth of brighter feathers; which, in turn, is used by the females as a quality assessing trait. Thus we assumed that higher the quality of the food supplied to the males, the greater would be the chances of their selection by the females. Accordingly, in the first experiment of food quality, females paired with the males that were fed with high quality diet.

Ptilochronology, a technique to quantify feather growth in terms of daily growth bars and which has been incorporated to study a diverse range of subjects (Grubb 1989), was used in our study. In the first experiment, where feed quality was manipulated, feather growth was unaffected. There is no any reference found in context to manipulation of diet. Thus this study provides first evidence that quality of food does not affect feather growth.

According to my hypothesis, I considered that feather growth bars are the predictors of mate choice decisions in females. This was contradicted in this experiment wherein the mate choice criterion of females was independent of the feather growth bar width (Table 4A.2, Graph 6A.3, Graph 6A.4). Experimental males selected by the females had narrower growth bars though fed with high quality diet. Thus, we must consider other factors that could influence growth bar width in experimental males. Grubb *et al.* (1991) discussed some other factors that influence feather growth rate such as age, sex, season, ambient temperature and photoperiod. However, because my experiment was done in controlled conditions and factors like temperature, photoperiod, season were maintained constant, it appears that age is the major contributing factor; which might have confounded this study.

Experiment 2–Photoperiod

Ball *et al*. (2008) discussed the twofold effects of the lengthening periods of daylight in temperate zone birds in spring on the reproductive physiology of birds (reference in Dawson *et al.,* 2001). Some are 'photo stimulated' birds which are reproductively active due to the stimulating effects of lengthening days whereas some are 'photo refractory', in which reproductive systems are regressed due to long days. Based on Dawson (2001), my results prove that Japanese Quail are photo refractory birds, in which they develop an inhibitory effect under extended light periods. This supports our results that sexual selection is the independent function of extended photoperiods.

White *et al*. (1992) found in American Tree Sparrows that the period of growth, but not the rate of growth, of induced feathers was greater under exposure to short periods of photoperiod. Also regrowth of contour feathers in Japanese Quail, was studied by Honda *et al*. (1982) who suggested that shortened daily photoperiods may have stimulated hormonal conditions characteristic of the period of fall molt. Grubb *et al*. (1991) considered day length to be one of the factors influencing feather growth in Northern Cardinals. Feather growth of Japanese Quails in our experiment too was dependent on photoperiod. Growth bar comprises of a pair of consecutive light and dark band, representing 24 hours of feather growth (Michener and Michener 1938). Out of which darker bands are derived from the material laid down in the follicle during the day and lighter bands consist of material laid down during the night (Wood 1950). Thus width of feather growth bars is dependent on the duration of light to which the birds were exposed during feather growth.

In this photoperiod experiment of photoperiod, females preferred to mate with the males with wider growth bars independent of the light and dark duration of 17:7 and 13:11 (Table 4B.2, Graph 6B.3, Graph 6B.4, Graph 6B.5, Graph 6B.6). In both sets of experiments, in shorter as well as longer light duration, males selected by the females had wide growth bars as compared to the experimental males. As growth bar widths are the indicators of individual quality, they also reflect its ability to maintain good body condition throughout the year (Takaki *et al.,* 2001). This supports my findings that females consider feather growth bar width while selecting their mate.

Experiment 3–Food Quantity

Soares *et al.* (2003), recommended 23.08 per cent and 21.95 per cent crude protein diet for Japanese Quail because there is no effect of protein levels on feed intake and feed conversion, *i.e.*, the morphology of the quails is not affected. This might be the reason that sexual selection in quails remained unaffected by the variation in quantity of protein. Shrivastav and Panda (1999) recommended 27 per cent protein diet for quails up to 2 weeks of age and 24 per cent protein diet for quails from 3 to 5 weeks of age. We tested the hypothesis for levels as low as 16 per cent and 18 per cent protein diets and which are lower than the recommended protein diets but sexual selection decisions of the females remained unaffected.

White and Kennedy (1992) suggested that the duration of rectrix growth, measured in days, is controlled hormonally while the rate of growth, measured by growth bar width, is controlled nutritionally in American Tree Sparrows. Jenkins (2001) found a significant effect of enriched (high protein) diet on feather growth in White-throated Sparrows, *Zonotrichia albicollis*. Similar to this, our results suggested the dependence of high protein contents (18 per cent) in the diet on ptilochronology of the bird, but feather growth remained unaffected under low protein diet (16 per cent). Thus, it appears that feather growth rate is affected when protein content in the diet is higher.

In both sets of experiments (16 per cent and 18 per cent protein diet), females paired with the males which had wider growth bar widths (Table 4C.2, Graph 6C.3, Graph 6C.4, Graph 6C.5, Graph 6C.6). Concurring to the findings of Takaki *et al.* (2001), my results too found growth bar width to be an effective indicator of individual quality.

Experiment 4–Ambient Temperature

A recent study on Great Tits showed that ambient temperature had a causal effect on timing of reproduction (Visser *et al.*, 2009). Ambient temperature is involved in the mechanism controlling annual reproductive cycle in Japanese Quail (Wada *et al.*, 1990). Ambient temperature also significantly affected the photoperiodically induced gonadal development and regression in mountain White-crowned Sparrow, *Zonotrichia leucophrys oriantha* (Wingfield *et al.*, 2003). Day-to-day variation in temperature, as well as mean temperature, affects energy expenditure, which have consequences for egg production in Japanese Quail (Pendlebury *et al.*, 2004). Carotenoids are proved to be a limiting resource during periods of cold exposure and that they might be prioritized for self-maintenance at the expense of the ornament in male Zebra Finch, *Taeniopygia guttata* (Eraud *et al.*, 2007). Temperature also influenced the breeding phenology of arid species such as the Trumpeter Finch, *Bucanetes githagineus* (Barrientos *et al.*, 2007). These studies revealed that temperature plays an important role in avian lifecycle including their morphology, phenology, embryology and most importantly their breeding biology. In parallel to all these studies our study has also demonstrated a significant effect of ambient temperature on sexual selection in Japanese Quail.

The ptilochronology in the fourth experiment of ambient temperature revealed that the feather growth was an independent function of temperature. This concurs with Zuberbier and Grubb (1992) who found no significant effect of either wind, temperature or their interaction on the rate of induced rectrices of White-breasted Nuthatches maintained on ad libitum food.

In our experiment, males reared in relatively cool temperatures were selected by the females for pairing. These males had wider growth bars as compared to the males exposed to higher temperature (Table 4D.2, Graph 6D.3, Graph 6D.4). Thus again concurring with the findings of Takaki *et al.* (2001), growth bar width were found to be an effective measure while taking mate choice decisions by females.

Chapter 7
Conclusion

This study provides a different aspect of sexual selection in relation to the daily growth rate of tail feathers. The sexual selection decisions of the females were compared with the comparative average growth bar widths of control and experimental birds. The results of three experiments of photoperiod, food quantity and ambient temperature showed that feather growth rate plays an important role in making mate choice decisions by females, whereas only the results of first experiment had contradicted the hypothesis that sexual selection does not depend on feather growth rate in males.

Thus the results support our hypothesis that feather growth bar width affects mate choice in Japanese Quail. Feather growth rate is proved to be a reliable indicator of male quality during female choice, and allows us to better understand the sexual selection in Japanese Quail. Thus, ptilochronology has opened an additional door into its application to study a new and different aspect of sexual selection.

Chapter 8
Abnormal Behavior

During the study period, a strikingly different behavior of the individually caged experimental birds was observed occasionally.

The males used for the experiments were kept isolated according to the need of the experiment right from the age when they attained sexually matured (3-4 weeks). They habituated to the conditions provided. But after 4-5 weeks when selection experiments were performed and females exposed to them, they got sexually excited. This sexual excitement was observed to varying degrees in most of the birds. These birds mounted on the drinker and feeder and release their semen.

This type of behavior was observed in isolated caged birds which do not get sexual partner for mating. Thus they developed abnormal behavior and expressed their frustrations on other objects either a feeder or drinkers available in the cage (Falk 1977; Heil *et al.*, 2007).

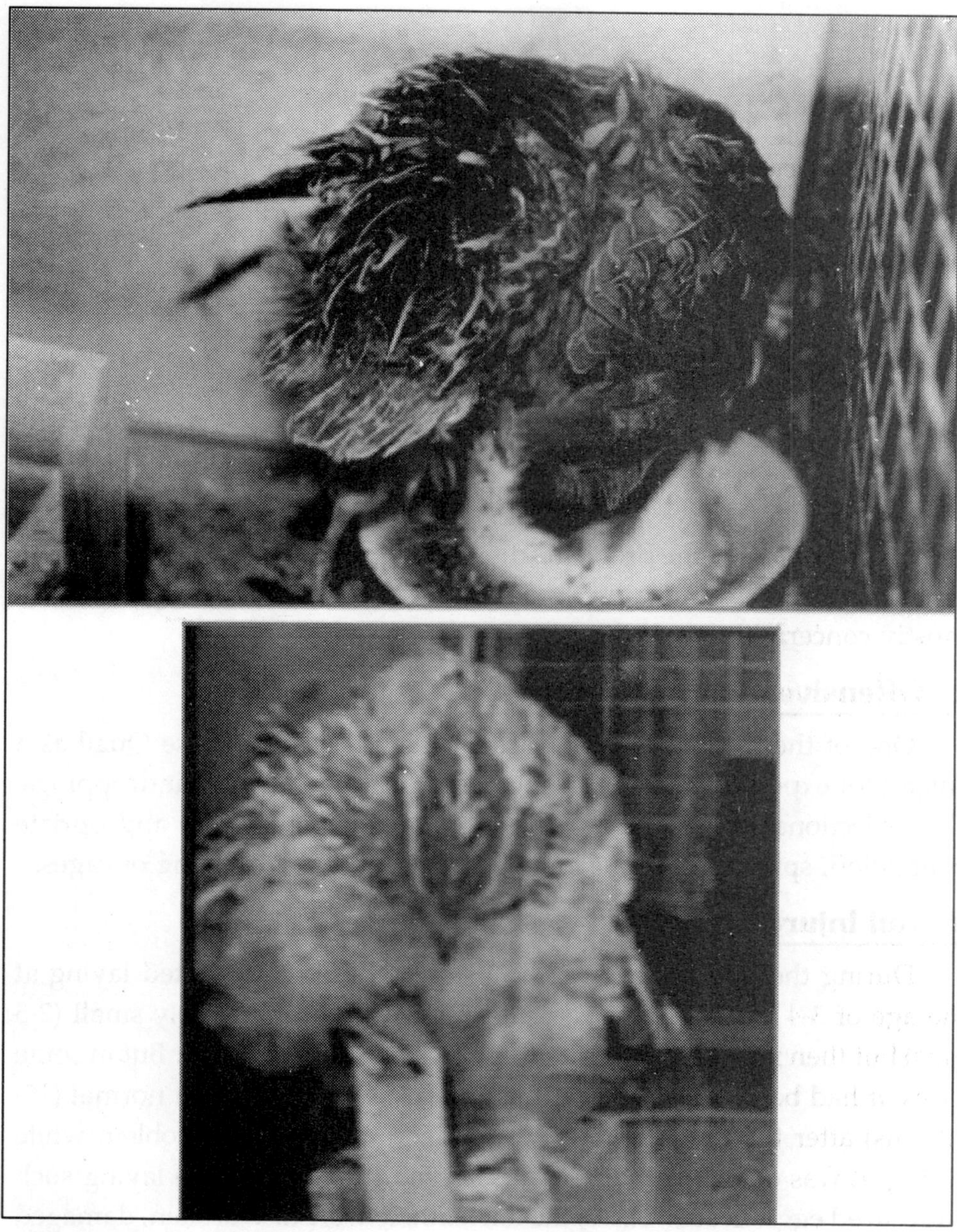

Plate 26: Aberrant Behaviour.
Figure: Isolated individual male Japanese Quail mounting on drinker and feeder.

Chapter 9
Obstacles

Since experiments were done in controlled conditions, we managed to overcome most of the problems discussed in Chapter-1. These and some other obstacles faced during study are discussed in this chapter. These mostly concern birds which were caged in groups.

1. Offensive Odour of Droppings

One of the major problems faced while rearing Japanese Quail as a subject for experiments is the strong odour of ammonia in their droppings. This objectionable smell was reduced to a negligible level by appropriate ventilation, spreading of hygienic sawdust and daily cleaning of cages.

2. Tail Injury

During the initial study period (Sep 2009), females started laying at the age of 3-4 weeks. Egg size of first egg was comparatively small (2-5 gms) but then increased and remained consistent (10-12 gms). But in some cases, it had been observed that the egg size was larger than normal (15-18 gms) after the first egg. Such large sized egg, created problem while laying. It was observed that females injured their vent while laying such large sized eggs. In such cases, the tail feathers were also broken, damaged and blood oozed out from the injured portion.

Such injured female was then caged separately to avoid any further injury in the same cage. Proper medical care was accorded to these females.

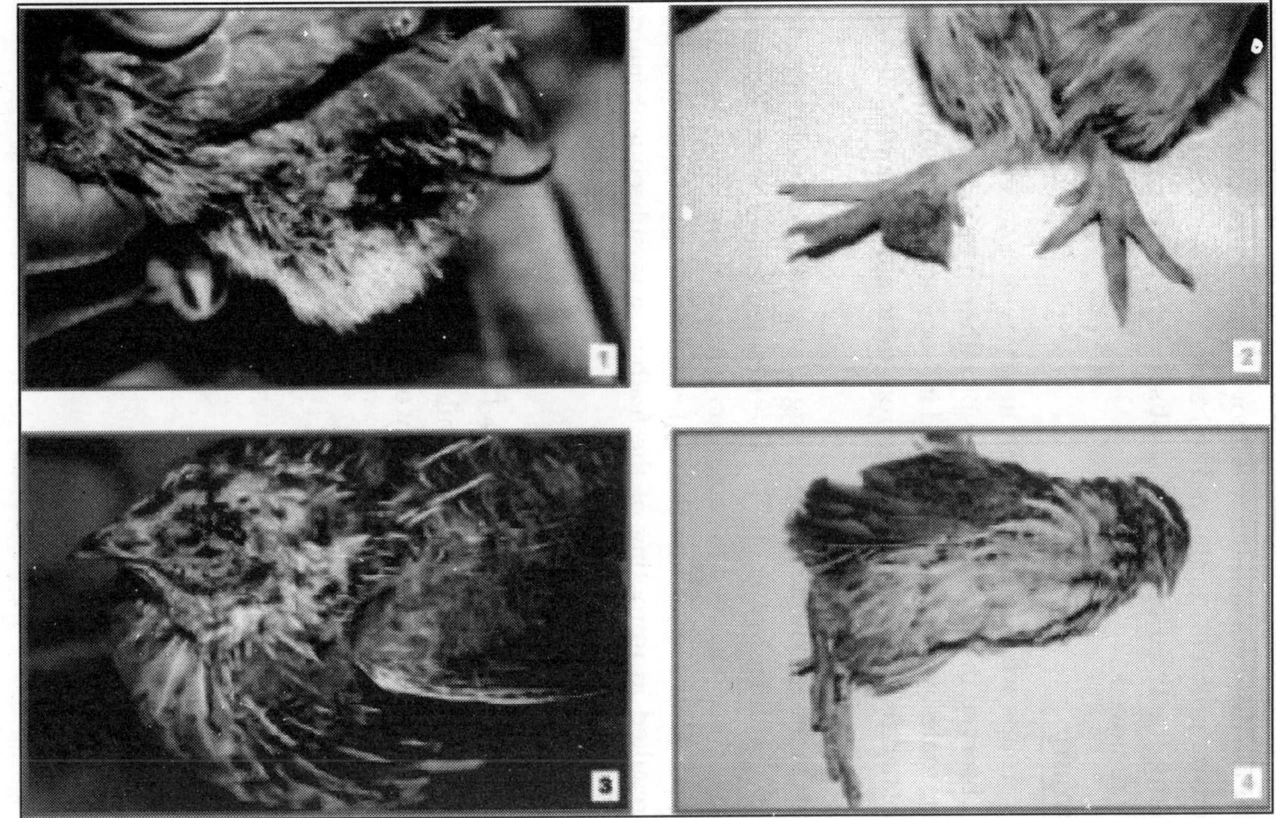

Plate 27: Obstacles.
Figure 1: Injured tail of female Japanese Quail due to laying of large sized egg; Figure 2: Solid structure formed due to wet floor; Figure 3: Head inury caused due to aggressive homosexual mountings in male Japanese Quails; Figure 4: Death caused during acclimatization of the birds.

3. Solid Structures in Hind Limbs

Formation of solid keratinized structures was observed in the hind limbs of those birds caged in groups. While moving in the cage, they knocked over the drinkers and the wet floor induces these solid structures which can create problems if allowed to harden. These were avoided by regular changing the floor. And if observed, were carefully and immediately removed.

4. Head Injury

Birds in groups were found with an injury with a severe head. This was mostly observed in males kept in groups rather in grouped females. It appears these males were that participated in homosexual mountings and that were isolated from the females right from their age of attaining sexual maturity.

Japanese Quails are known for their aggressive behavior (Selinger and Bermant 1967, Schlinger *et al.,* 1987). The isolated males when became sexually active, were observed to mount on other males in the same cage which objected. However, when the males tried to forcibly mount the aggression resulted in uprooting of the feathers and subsequent injury (Plate 27).

5. Mortality

During the study period, we evidenced bird mortality. Normally, 1-2 per cent is the normal mortality rate while acclimatizing in a different environment. But other than this, mortality occurred mostly due to laying.

Appendix–1

SCIENTIFIC NAMES OF BIRD SPECIES MENTIONED

Sl.No.	Common Name	Scientific Name
1.	American Kestrel	*Falco sparverius*
2.	American Tree Sparrow	*Spizella arborea*
3.	Antarctic Cormorant	*Phalacrocorax bransfieldensis*
4.	Bearded Vulture	*Gypaetus barbatus*
5.	Black Wheatear	*Oenenthe leucura*
6.	Blue-black Grassquit	*Volatinia jacarina*
7.	Blue-black Grosbeak	*Guiraca caerulea*
8.	Brown-headed Cowbird	*Molothrus ater*
9.	California Quail	*Callipepla californica*
10.	Canarie	*Serinus canaria*
11.	Carolina Chickadee	*Parus carolinensis*
12.	Common Yellowthroat	*Geothlypis trichas*
13.	Dark-eyed Juncos	*Junco hyemalis*
14.	Desert Finch	*Rhodospiza obsoleta*

Sl.No.	Common Name	Scientific Name
15.	Domestic Chicken/Fowl	*Gallus gallus domesticus*
16.	Downy Woodpecker	*Picoides pubescens*
17.	Eastern Bluebird	*Sialia sialis*
18.	Florida Scrub-Jay	*Aphelocoma caerulescens*
19.	Graceful Prinia/Warbler	*Prinia gracilis*
20.	Gray Jay	*Perisoreus canadensis*
21.	Great Cormorant	*Phalacocorax carbohanedae*
22.	Great Tit	*Parus major*
23.	Greylag Geese	*Ancer ancer*
24.	Guppies	*Poecilia reticulata*
25.	House Finch	*Carpodacus mexicanus*
26.	Jackson's Widow bird	*Euplectes jacksonia*
27.	Japanese Quail	*Coturnix coturnix japonica*
28.	Laysan Albatross	*Phoebastria immutabilis*
29.	Lesser Snow Goose	*Chen c. caerulescens*
30.	Loggerhead Shrike	*Lanius ludovicianus*
31.	Northern Cardinal	*Cardinalis cardinalis*
32.	Orange-fronted Parakeet	*Aratinga canicularis*
33.	Pied Flycatcher	*Ficedula hypoleuca*
34.	Red-collared Widow bird	*Euplectes ardens*
35.	Red Jungle Fowl	*Gallus gallus*
36.	Sage Grouse	*Centrocurcus urophasianus*
37.	Siberian Jay	*Perisoreus infaustus*
38.	Southern Giant Petrel	*Macronectes giganteus*
39.	Styan's Grasshopper Warbler	*Locustella pleskei*
40.	Tree Swallow	*Tachycineta bicolor*
41.	Trumpeter Finch	*Bucanetes githagineus*

Sl.No.	Common Name	Scientific Name
42.	Tufted Titmouse	*Parus bicolor*
43.	Upland Buzzard	*Buteo hemilasius*
44.	White-backed Woodpecker	*Dendrocopos leucotos*
45.	White-breasted Nuthatch	*Sitta carolinensis*
46.	White-crowned Sparrow	*Zonotrichia leucophrys*
47.	White-throated Sparrow	*Zonotrichia albicollis*
48.	White-winged Becard	*Pachyramphus polychopterus*
49.	Zebra Finch	*Taeniopygia guttata*

Appendix–2

LIST OF ABBREVIATIONS

Sl.No.	Abbreviation	
1.	Expt 1–FQ	Experiment 1–Food Quality
2.	Expt 2a–P	Experiment 2a–Photoperiod (L:D=17:7)
3.	Expt 2b–P	Experiment 2b–Photoperiod (L:D=13:11)
4.	Expt 3a–FQt	Experiment 3a–Food Quantity (18 per cent protein)
5.	Expt 3b–FQt	Experiment 3b–Photoperiod (16 per cent protein)
6.	Expt 4–AT	Experiment 4–Ambient Temperature

References

Alantalo R., Lundberg A., and Stahlbrandt K. (1984) – Female mate choice in the Pied flycatcher *Ficedula hypoleuca. Behavioral Ecology and Sociobiology,* 14(4): 253-261.

Andersson S. (1989) – Sexual selection and cues for female choice in leks of Jackson's widowbird *Euplectes jacksonia,* 25(6): 403-410.

Bagemihl B. (1999) – Biological exuberance, Animal homosexuality and natural diversity. *St. Martin's Press,* New York, 1-750.

Ball G. and Ketterson E. (2008) – Sex differences in the response to environmental cues regulating seasonal reproduction in birds. *Philosophical Transactions of Royal Society, 363*: 231-246.

Ballentine B. and Hill G. (2003) – Female mate choice in relation to structural plumage coloration in Blue Grosbeaks. *The Condor,* 105: 593-598.

Bertran J. and Margalida A. (2003) – Male-male mountings in polyandrous Bearded Vultures *Gypaetus barbatus*: an unusual behavior in raptors. *Journal of Avian Biology,* 34: 334-338.

Blanco G., Dawson R. and Murza G. (2002) – Bill abnormalities in a pair of Black Wheatears *Oenanthe leucura. Butll. GCA,* 43-46.

Bortolotti G., Dawson R. and Murza G. (2002) – Stress during feather development predicts fitness potential. *Journal of Animal Ecology,* 71: 333-342.

Bossema I. and Roemers E. (1984) – Mating strategy, including mate choice, in Mallards. *Ardea,* 73: 147-157.

Bostrom M. and Ritchison G. (2006) – Possible relationships between morphology, territory quality, and skin color of American Kestrels. *Journal of Field Ornithology,* 77(4): 392-398.

Buchanan O. (1966) – Homosexual behavior in wild Orange-fronted Parakeets. *The Condor,* 68(4): 399-400.

Cain J. and Cawley W. (1914) – Japanese Quail (*Coturnix*). *The Texas Agricultural experiment Station and the Texas Agricultural Extension Service,* 2M: 6-79.

Calkins J. and Burley N. (2003) – Mate choice for multiple ornaments in the California Quail, *Callipepla californica. Animal Behavior,* 65: 69-81.

Candolin U. (2003) – The use of multiple cues in mate choice. *Biological Reviews,* 78: 575-595.

Carlson A. (1998) – Territory quality and feather growth in the White-backed Woodpecker, *Dendrocopos leucotos. Journal of Avian Biology,* 29: 205-207.

Carrascal L., Senar J., Mozetich I., Uribe F. and Domenech J. (1998) – Interactions among environmental stress, body condition, nutritional status, and dominance in Great tits during winter. *Auk,* 115: 727-738.

Casaux R. (2004) – Beak deformation in an Antarctic Cormorant *Phalacrocorax [Atriceps] Bransfieldensis* chick. *Marine Ornithology,* 32: 109-110.

Charde P. (1998) – Effect of pharmacological components on biology of *Coturnix coromendelica.* A Ph.D Thesis submitted to *Rashtrasant Tukdoji Maharaj Nagpur University,* Nagpur.

Conover M. and Hunt G., Jr. (1979) – Female-female pairings and sex ratios in gulls, A historical perspective. *Willson Bulletin,* 96: 619-625.

Craves J. (1994) – Passerines with deformed bills. *North American Bird Bander,* 19(1): 14-19.

Crawford L., Holloway K. and Domjan M. (1993) – The nature of sexual reinforcement. *Journal of the Experimental Analysis of Behavior*, 60: 55-66.

Dawson A., King V., Bentley G. and Ball G. (2001) – Photoperiodic control of seasonality in birds. *Journal of Biological Rhythms*, 78: 365-380.

Dolenec Z., Mrakovcic M. and Delic A. (2005) – Egg dimensions of the Great tit (*Parus major*) in croatia. *Polish Journal of Ecology*, 53(1): 143-145.

Doucet S. (2002) – Structural plumage coloration, male body size, and condition in the Blue-black Grassquit. *The Condor*, 104: 30-38.

Easterling D., Horton B., Jones P., Peterson T., Karl T., Parker D., Salinger J., Razuvayev V., Plummer N., Jamason P. and Folland C. (1997) – Maximum and minimum temperature trends for the globe. *Science*, 277: 364-367.

Encabo S., Monros J. and Barba E. (2001) – Egg size variation in a Mediterranean Great tit *Parus major* population. *Ardeola*, 48(1): 68-70.

Eraud C., Devevey G., Gaillard M., Prost J., Sorci G. and Faivre B. (2007) – Environmental stress affects the expression of a carotenoid-based sexual trait in male zebra finches. *Journal of Experimental Biology*, 210: 3571-3578.

Ernst R. (1978) – Raising and Propogating Japanese Quail. Division of Agricultural Sciences, University of California, 2738: 1-8.

Falk J. (1977) – The origin and functions of adjunctive behavior. *Animal Learning and Behavior*, 5(4): 325-335.

Fitzpatrick S. (1998) – Colour schemes for birds: structural coloration and signals of quality in feathers. *Ann. Zool. Fennici*, 35: 67-77.

Freeman-Gallant C., Taff C., Morin D., Dunn., Whittingham L., Tsang S. (2009) – Sexual selection, multiple male ornaments, and age-and condition-dependent signaling in the common Yellowthroat. *Evolution*, 1-11. doi:10.1111/j.1558-5646.2009.00873.x.

Fukuda M. (1992) – Male-male pairing of the Great Cormorant (*Phalacocorax carbohanedae*). *Colonial Waterbird Society Bulletin,* 16: 62-63.

Gabhane R. (2011) – Ptilochronology – Feather study and biology of *Coturnix coturnix japonica* quail bird. A Ph.D Thesis submitted to *Rashtrasant Tukdoji Maharaj Nagpur University,* Nagpur.

Galef B., Jr. (2008) – Social influences on the mate choices of male and female Japanese Quail. *Comparitive Cognition and Behavior Reviews,* 3: 1-12.

Gibson R. and Bradbury J. (1986) – Male and female mating strategies on Sage Grouse (*Centrocurcus urophasianus*) leks. *Ecological Aspects of Social Evolution,* 379-398. Princeton: Princeton University Press.

Gienapp P. and Merila J. (2010) – Genetic and environmental effects on a condition-dependent trait: feather growth in Siberian Jays. *Journal of Evolutionary Biology,* 23: 715-723. (doi:10.1111/j.1420-9101.2010.01949.x)

Grafen A. (1990) – Biological signals as handicaps. *J. theor. Biol.,* 144: 517-546.

Grether G., Hudon J. and Millie D. (1999) – Carotenoid limitation of sexual coloration along an environmental gradient in Guppies. *The Royal Proceedings,* 266(1426): 1317-1322.

Grubb T., Jr. (1989) – Ptilochronology: Feather growth bars as indicators of nutritional status. *The Auk,* 106: 314-320.

Grubb T., Jr. (1991) – A deficient diet narrows growth bars on induced feathers. *The Auk,* 108: 725-727.

GrubbT., Jr. (2006) – Ptilochronology. Feather time and biology of birds. *Oxford: Oxford University Press.*

Grubb T., Jr. and Cimprich D. (1990) – Supplementary food improves the nutritional condition of wintering woodland birds: Evidence from Ptilochronology. *Ornis Scandinavica,* 21(4): 277-281.

Grubb T., Jr. and Yosef R. (1994) – Habitat-specific nutritional condition in Loggerhead Shrikes (*Lanius ludovicianus*): Evidence from Ptilochronology. *The Auk,* 111(3): 756-759.

GrubbT., Jr., Waite T. and Wiseman A. (1991) – Ptilochronology: Induced feather growth in Northern Cardinals varies with age, sex, ambient temperature, and day length. *Wilson Bulletin*, 103(3), 435-445.

Grubb T., Jr., Woolfenden G. and Fitzpatrick J. (1998) – Factors affecting nutritional condition of fledgling Florida Scrub-Jays: A Ptilochronology approach. *The Condor*, 100: 753-756.

Grubb T., Jr. (2006) – Ptilochronology-Feather time and Biology of the Birds. Oxford University Press.

Hagelin J. and Ligon D. (2001) – Female Quail prefer testosterone-mediated traits, rather than the ornate plumage of males. *Animal Behavior*, 61: 465-476.

Heil L., Fernandez-Juricic E., Renison D., Cingolani A. and Blumstein D. (2007) – Avian responses to tourism in the biogeographically isolated high Cordoba Mountains, Argentina. *Biodiversity Conservation*, 16: 1009-1026.

Heindl M. and Winkler H. (2003) – Female Canaries (*Serinus canaria*) associate more with males that contrast strongly against the background. *Ethology*, 109: 259-271.

Hill G. (1990) – Female House Finches prefer colorful males: sexual selection for a condition-dependent trait. *Animal Behavior*, 40: 563-572.

Hill G. (1991) – Plumage coloration is a sexually selected indicator of male quality. *Letters to Nature*, 350: 337-339.

Hill G. and Montgomerie R. (1994) – Plumage colour signals nutritional condition in the house finch. *Biological Sciences*, 258(1351): 47-52.

Holveck M. and Riebel K. (2009) – Low quality females prefer low-quality males when choosing a mate. *Proceedings of the Royal Society Biological Sciences*, 277: 153-160. doi:10.1098/rspb.2009.1222.

Honda H., Tanaka K., Minamino T. and Konishi T. (1982) – Control of contour feather growth of Japanese Quail. *The journal of experimental zoology*, 220: 311-319.

Horak P., Vellau H., Ots I. and Moller A. (2000) – Growth conditions affect carotenoid-based plumage coloration of great tit nestlings. *Naturwissenschaften*,87: 460-464.

Hoyt D. (1979) – Practical methods of estimating volume and fresh weight of bird eggs. *The Auk*, 96: 73-77.

Huber R. and Martys M. (1993) – Male-male pairs in Greylag Geese (*Ancer ancer*). *Journal of Ornithology*, 134: 155-164.

Jenkins K., Hawley D., Farabaugh C. and Cristol D. (2001) – Ptilochronology reveals differences in condition of captive White-throated Sparrows. *The Condor*, 103(3): 579-586.

Keyser A. and Hill G. (2000) – Structurally based plumage coloration is an honest signal of quality in male Blue Grosbeaks. *Behavioral Ecology*, 11(2): 202-209.

Leonard M. and Zanette L. (1998) – Female mate choice and male behavior in domestic Fowl. *Animal Behavior*, 56: 1099-1105.

Ligon J. and Zwartjes P. (1995) – Ornate plumage of male Red Junglefowl does not influence mate choice by females. *AnimalBehavior*, 49: 117-125.

Ligon J., Kimball R. and Zwartjes M. (1998) – Mate choice by female Red Junglefowl: the issues of multiple ornaments and fluctuating asymmetry. *Animal Behavior*, 55: 41-50.

Lombardo M., Bosman R., Faro C., and Houtteman S. (1994) – Homosexual copulations by male Tree Swallows. *The Wilson Bulletin*, 106(3): 555-557.

Machmer M., Esselink H., Steeger C. and Ydenberg R. (1991) – The occurrence of fault bars in the plumage of nestling Ospreys. *Ardea*, 80: 261-272.

Mahler B., Araujo L. and Tubaro P. (2003) – Dietary and sexual correlates of carotenoid pigment expression in dove plumage. *The Condor*, 105: 258-267.

Marti L., Bellagamba P., and Coria N. (2008) – Beak deformation in a southern giant petrel *Macronectes giganteus* chick. *Marine Ornithology*, 36: 195-196.

Matysiokova B. and Remes V. (2010) – Assessing the usefulness of ptilochronology in the study of melanin- and carotenoid-based ornaments in the Great Tit (*Parus major*). *Ibis*, 152: 397-401.

McGlothin J., Duffy D., Henry-Freeman J. and Ketterson E. (2007) – Diet quality affects an attractive white plumage pattern in Dark-eyed Juncos (*Junco hyemalis*). *Behavior Ecology and Sociobiology,* 61: 1391-1399.

McGraw and Hill (1999) – Induced homosexual behavior in male House Finches (*Carpodacus mexicanus*): the "Prisoner Effect". *Ethology Ecology and Evolution,* 11: 197-201.

Michener H. and Michener J. (1938) – Bars in flight feathers. *Condor,* 40: 149-160.

Mills A., Crawford L., Domjan M. and Faure J. (1997) – The behavior of the Japanese or domestic Quail *Coturnix japonica. Neuroscience and Behavioral Reviews,* 21(3): 261-281.

Minvielle F. (2004) – The future of Japanese Quail for research and production. *World's Poultry Science Journal,* 60: 500-507.

Moller A. (1989) – Viability costs of male tail ornaments in a Swallow. *Letters to Nature,* 339: 132-135.

Moreno J. and Osorno J. (2003) – Avian egg color and sexual selection: does eggshell pigmentation reflect female condition and genetic quality? *Ecology Letters,* 6: 803-806.

Nager R. and Zandt H. (1994) – Variation in egg size in Great tits. *Ardea.* 82: 315-328.

Naoroji R. (2007) – Birds of prey of the Indian subcontinent. Om Books International.

Ophir A. and Galef B. (2003) – Female Japanese Quail affiliate with live males that they have seen mate on video. *Animal Behavior,* 66: 369-375.

Pendlebury C., MacLeod M. and Bryant D. (2004) – Variation in temperature increases the cost of living in birds. *The Journal of Experimental Biology,* 207: 2065-2070.

Persaud K. and Galef B. Jr. (2005) – Female Japanese quail (*Coturnix japonica*) mated with males that harassed them are unlikely to lay fertilized eggs. *Journal of Comparative Psychology,* 119(4): 440–446.

Potti J. (1993) – Environmental, ontogenic, and genetic variation in egg size of Pied Flycatchers. *Can. J. Zool.*, 71: 1534-1542.

Pryke S., Andersson S. and Lawes M. (2001) – Sexual selection of multiple handicaps in the red-collared Widowbird: female choice of tail length but not carotenoid display. *Evolution* 55(7): 1452–1463.

Quinn T., Davies J., Cooke F. and White B. (1989) – Genetic analysis of off-spring of a female-female pair in the Lesser Snow Goose (*Chen c. caerulescens*). *Auk,* 106:177-184.

Randall M. and Bolla G. (2008) – Raising Japanese Quail. *Primefact,* 602 second edition.

Regan E. and Krakauer (2000) – Removal of adult males from the rearing environment increases preference for same-sex partners in the zebra finch. *Animal Behavior,* 60: 47-53.

Rintoul D. (2005) – Beak deformity in a brown-headed cowbird, with notes on causes of beak deformities in birds. *Kansas Ornithological Society Bulletin,* 56: 29-32.

Sanford J. (1957) – A progress report of Coturnix Quail Investigations in Missouri. *Proceedings of North American Wildlife Conference,* 22: 316-359.

Schlinger B., Palter B. and Callard G. (1987) – A method to quantify aggressiveness in Japanese Quail (*Coturnix coturnix japonica*). *Physiology and Behavior,* 40 (3): 343-348.

Selinger H. and Bermant G. (1967) – Hormonal control of aggressive behavior in Japanese Quail (*Coturnix coturnix japonica*). *Behavior,* 28 (3/4): 255-268.

Senar J., Figuerola J. and Domenech J. (2003) – Plumage coloration and nutritional condition in the Great tit *Parus major*: the roles of carotenoids and melanins differ. *Naturwissenschaften,* 90: 234-237.

Serrano D. and Jovani R. (2005) – Adaptive fault bar distribution in a long-distance migratory, aerial forager passerine? *Biological Journal of the Linnean Society,* 85: 455-461.

Siefferman L. and Hill G. (2003) – Structural and melanin coloration indicate parental effort and reproductive success in male Eastern Bluebirds. *Behavioral Ecology,* 14(6): 855-861.

Siefferman L. and Hill G. (2005a) – Evidence for sexual selection on structural plumage coloration in female Eastern Bluebirds (*Sialia sialis*). *Evolution,* 59(8): 1819-1828.

Siefferman L. and Hill G. (2005b) – Male Eastern Blue birds trade future ornamentation for current reproductive investment. *Biological Letters,* 1: 208-211. doi: 0.1098/rsbl.2004.0274

Siefferman L. and Hill G. (2007) – The effect of rearing environment on blue structural coloration of Eastern Blue birds (*Sialia sialis*). *Behavioral Ecology and Sociobiology,* 61: 1839-1846.

Siefferman L., Shawkey M., Bowman R. and Woolfenden G. (2008) – Juvenile coloration of Florida Scrub-Jays (*Aphelocoma coerulescens*) is sexually dichromatic and correlated with condition. *Journal of Ornithology,* 149: 357-363.

Soares R., Fonseca J., Santos A. and Mercandante M. (2003) – Protein requirement of Japanese Quail (*Coturnix coturnix japonica*) during rearing and laying periods. *Revista Brasileira de Ciência Avícola,* 5(2). (doi: 10.1590/S1516-635X2003000200010)

Srivastava S. and Srivastava S. (2003) – Fundamentals of Statistics. *Anmol Publications.*

Sundev G., Yosef R. and Birazana O. (2009) – Brandt's vole density affects nutritional condition of Upland Buzzard *Buteo hemilasius* on the Mongolian grassland steppe. *Ornis Fennica,* 86: 131-139.

Takaki Y., Eguchi K. and Nagata H. (2001) – The growth bars on tail feathers in the male Styan's grasshopper warbler may indicate quality. *Journal of Avian Biology,* 32(4): 319-325.

Vali N. **(2008)** – The Japanese Quail: A Review. *International Journal of Poultry Science,* 7(9): 925-931.

Vasconcelos M. and Rodrigues M. (2006) – Bill deformity in a White-winged Becard, *Pachyramphus polychopterus,* (Aves: Suboscines: Tityridae) from Minas Gerais, Brazil. *Revista Brazileira de Ornitologia,* 14(2): 165-166.

Visser M., Holleman L. and Caro S. (2009) – Temperature has a causal effect on avian timing of reproduction. *Proceedings of the royal Society,* 276: 2323-2331.

Wahab H. (2002) – Quails could reduce protein deficiency in poor countries. *Elsevier*, 18(6): 02.

Waite T. (1990) – Effects of caching supplemental food on induced feather regeneration in wintering Gray Jays *Perisoreus canadensis:* a ptilochronology study. Ornis Scandinavica, 21: 122-128.

White D. and Galef B., Jr. (2000) – Differences between the sexes in direction and duration of response to seeing a potential sex partner mate with another. *Animal Behaviour*, 59: 1235-1240.

White D. and Kennedy E. (1992) – Growth of induced feathers in photo stimulated American tree Sparrows. *The Condor*, 94: 543-545.

Wood H. (1950) – Growth bars in feathers. *Auk*, 67: 486-491.

Woodard A., Abplanalp H., Wilson W. and Vohra P. (1973) – Japanese Quail husbandary in the laboratory. *Department of Avian Sciences, University of California*, 1-22.

Yosef R. (1997) – On habitat-specific nutritional condition in Graceful warblers *Prinia gracilis*: Evidence from Ptilochronology. *Journal fur Ornithologie*, 309-313.

Yosef R. and Grubb T. (1992) – Territory size influences nutritional condition in nonbreeding Loggerhead Shrikes (*Lanius ludovicianus*): A Ptilochronology approach. *Conservation Biology*, 6(3): 447-449.

Yosef R. and Zduniak P. (2004) – Within-clutch variation in egg dimensions of Loggerhead shrike (*Lanius ludovicianus*) in south-central Florida. *Biological Letters*, 41(2): 155-162.

Yosef R. and Zduniak P. (2008) – Variation in clutch size, egg size variability and reproductive output in Desert Finch (*Rhodospiza obsoleta). Journal of Arid Environments*, 72: 1631-1635.

Young L., Zaun B. and VanderWerf E. (2008) – Successful same-sex pairing in Laysan albatross. *Biology letters (Animal Behavior)*, 4: 323-325.

Zahavi A. (1975) - Mate Selection – A selection for a handicap. *J. theor. Biol.*, 53: 205-214.

Zuberbier G., Grubb T., Jr. (1992) – Ptilochronology: Wind and cold temperatures fail to slow induced feather growth in captive White-breasted Nuthatches maintained on ad libitum food. *Ornis Scandinavica,* 23: 139-142.

Zuk M. and Bailey N. (2008) – Birds gone wild: same-sex parenting in albatross. *Trends in Ecology and Evolution,* 23(12): 658-660.

Internet Links:

1. http://www.icar.org.in/cari/quail.html
2. http://www.pjbs.org/ijps/fin1218.pdf
3. http://www.vethelplineindia.com/art_quails.htm
4. http://www.worldpoultry.net/article-database/quails-could-reduce-protein-deficiency-in-poor-countries-id1397.html

Index